High and Low Politics: Information Resources for the 80s

LIKE MATERIALS AND ENERGY . . .

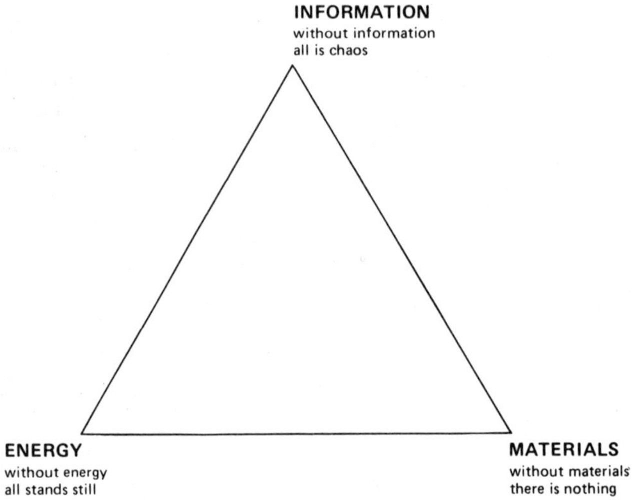

. . . INFORMATION IS A BASIC RESOURCE.

MODES	PRE-INDUSTRIAL	INDUSTRIAL	POST-INDUSTRIAL
Transforming Resource	Natural Power *Wind, Water, Draft animals, Human muscle*	Created Energy *Electricity — oil, gas, coal* *Nuclear power*	Information *Computer and data-transmission systems*
Strategic Resource	Raw Materials	Financial Capital	Knowledge

Source: Bell, Daniel. "Welcome to the Post-Industrial Society," *Physics Today*, Vol. 29, No. 2, February 1976, p. 47.

High and Low Politics: Information Resources for the 80s

Anthony G. Oettinger
Paul J. Berman
William H. Read

Ballinger Publishing Company • Cambridge, Massachusetts
A Subsidiary of J.B. Lippincott Company

This book is printed on recycled paper.

International Standard Book Number: 0–88410–064–2

Library of Congress Catalog Card Number: 77–1685

Printed in the United States of America

Library of Congress Cataloging in Publication Data

Oettinger, Anthony G
High and low politics.

Includes bibliographical references.
1. Telecommunication—United States—Addresses, essays, lectures.
I. Berman, Paul J., joint author. II. Read, William H., joint author.
III. Title.
HE7777.034 384'.0973 77–1685
ISBN 0–88410–064–2

Contents

List of Figures

Acknowledgments

Elizabeth Lazarus did spade work for *The Medium and the Telephone* during a summer internship in the Washington office of Representative Michael J. Harrington and through an independent study course at Harvard. Carol Weinhaus most ably traced and condensed voluminous but often elusive data. The following assisted in drafting *Computer or Communications?*: Susan Berman, Claire Bishop, Peter Hankin, Fred Hapgood, Susan Morrison, Andrew Reese, Carol Weinhaus, and Nikki Zapol.

Weinhaus and Bishop supervised the production of all three parts by Claire Bishop, Sylvia Cashton, Renate D'Arcangelo, Armand Dionne, William Minty, and Karin Young.

Special thanks are due to the following persons, who reviewed the study plans, supplied data, criticized drafts, or otherwise helped us in preparing the three parts of this book. However, they are not responsible for—nor do they necessarily agree with—the views expressed herein; any factual errors or misinterpretations are solely ours.

Elizabeth Allison
Robert Angliss
Harry Baer
William Baker
Robert Bennis
Hendrik Bode
Kurt Borchardt
Anne Branscomb
Stephen Breyer
Harvey Brooks
Joe Burke
Frank Burrows
William Capron
Abram Chayes
John Clippinger
William Comanor
Larry Darby
James Darr

Robert Dickinson
Wilson Dizard
Richard Etzi
Ned Farinholt
William Fishman
Ed Fitzgerald
Richard Freund
Richard Gabel
Joe Gibson
Dean Gillette
John Griffith
Wreatham Gathwright
Donald Hollis
Roland Homet
Gordon Imrie
Louis Jaffe
Charles Joyce
Erwin Krasnow
John LeGates
Elmer Lower
Preston Mabon
Gene Mater
William McGowan
Vincent Mosco
Delbert Myren
Joseph Nye
Ann Orlov
Daniel Parker
Tom Parmenter
James Peacock
Donald Peterson
Gerard Piel
Daniel Prives
Ithiel de Sola Pool
Gordon Raisbeck
John Richardson
Eugene Rostow
John Rothman
Peter Salisbury
Charles Sanders
David Seipp
Peter Shapiro
James Sichter
Ted Simis
Michael Spence
Richard Stewart
Bernard Strassburg
Neil Swift
Anthony Terracciano
J. Gordon Walter
Fred Warden
Earle Whitney
Nancy Wiatt
Wallace Wormley

Introduction

Information, as Harvard sociologist Daniel Bell has pointed out, is the transforming and strategic resource of American society today.

Since the early 1960s, the dominant trend in the U.S. labor force has been the growth of information occupations. Not only has there been an expansion in traditional areas such as education, but new occupations have come into being in fields such as computer programming. By the mid 1970s, these information occupations were employing close to half the American labor force. During the same period, the information industries, from broadcast television to legal services, also were on the rise, as information resources became increasingly essential for people to participate in America's advancing society. The information sector of the U.S. economy has already accounted for at least 20 percent of the U.S. gross national product, and it has been forecast to expand rapidly during the final quarter of the twentieth century.

What does this signify for America? A case in point is the doubling every two to three years in the number of overseas telephone calls. Information that flows via communications satellites and transistorized undersea cables has enabled transnational organizations to pursue global strategies—and thereby promote interdependence: economic . . . social . . . and political. Since telecommunications, like other information resources, is highly sensitive to government regulation and (even government ownership), control over telecommunications through government policies is a powerful lever on international relations.

The lifeline dependence of international banks on telecommunications illustrates America's new stakes in and issues associated with information resources. International banking has grown enormously. One indicator of this is the proliferation of American branch banks abroad, from 200 in 1965 to over 700 a decade later. Amendments to the Federal Reserve Act are thought of as typical policy tools for having influenced the spread of U.S. international banking. But policies like those governing communications satellites also affect the vital flow of money. And money, after all, is nothing but an information symbol that can be communicated electronically.

Information resources have become more important than ever to banking. Among the reasons: electronic financial transactions became increasingly common; prudent money management, international as well as domestic, has required rapid electronic communication between banks' headquarters and branches; and banking's modern version of Rothschild's couriers is its own private, global, electronic communications networks. Besides handling thousands of international monetary transactions daily, they also transmit economic intelligence.

The proliferation of American branch banks abroad mirrors the rise in influence of international banks in world affairs. Transnational banks cut costs and, more important, muster their strength through command and control of their branches. They do this through telecommunications networks. But while banks have acquired communications networks that resemble the U.S. State Department's "cablegram" system, banks have no diplomatic immunity. Thus their networks are a source of vulnerability and strategic weakness. In fact, banking networks are as vulnerable to disruptions as the Washington-Moscow "hotline"—which has been accidentally cut several times, once by a Finnish farmer plowing his field.

With so much at stake, it is not difficult to understand why the Society for Worldwide Financial Telecommunications (SWIFT) was formed in 1973. Following its creation, a series of crucial meetings was held that resulted in a decision that protects banking's lifelines by setting up the SWIFT network to provide high technical and political stability. Starting in 1976, SWIFT's computer/communications circuits became international lifelines for over 250 banks in North America and Western Europe. By 1978, the SWIFT network was expected to process some 300,000 financial transactions daily. SWIFT's lifelines were protected with duplication: even the vital twin switching centers were situated in two different countries—Belgium and the Netherlands—which are havens for information resources not unlike those politically stable and economically useful "free ports" for trade.

SWIFT may be but one new answer for safeguarding banking's international communications lifelines. Following the Pentagon's lead in setting up nuclear command and control systems, First National City Bank has considered experimenting with direct communications by satellite between Citibank's New York headquarters and its London branch. Yet, even as international banking by direct satellite appeared feasible, the United States faced a round of worrisome negotiations on the allocation of satellite frequencies at the International Telecommunications Union—another U.N. forum passing out of western control. So for banking, which in the computer age cannot return to couriers, the availability of frequencies for communications satellites looms as a "lifeline" issue.

International banking is not exceptional. National security relies on electronic sentries for intelligence gathering; truck drivers depend on citizen band radios to evade speed traps and obtain traffic information. In modern societies, information tends to be a practical, indeed a sensible source of power. This is not to say that the impact of information resources on America will diminish the importance of traditional concerns. As with the evolution from an agrarian to an industrial society, the newer does not replace, but rather adds to, the older.

As the United States moves into an "information society," how well is it prepared to understand the stakes and to address the issues of information resources? In 1972 the Program on Information Resources Policy (formerly Program on Information Technologies and Public Policy) was organized at Harvard University to provide policy perspectives on information resources. The Harvard Program's central concept is that information is just as much a basic resource as are materials and energy. Within that intellectual framework, the Program has sought to provide objective studies that describe areas of conflict, characterize significant players, enumerate stakes, identify forces in action, mark trends, discuss strategies, and assess policy options.

This book contains three of the Program's major policy studies. The reader will find some common threads running through each. One is the breaking down of differences between information technologies. Another is a lowering of barriers that once distinguished information industries. A third is new ways of acquiring and using information.

The trilogy begins with *The Medium and the Telephone: The Politics of Information Resources.* While he was a student at the Harvard Law School, Paul J. Berman worked with Professor Anthony G. Oettinger, the Program Director, on a project that traced the politics, economics, and technology of telecommunications pricing and costing policy, using newspaper and television news as a concrete example for probing policy impacts. They examined divided federal-state jurisdiction, renewed competition in telecommunications, rate setting mechanisms, and technological change. Players were identified as well as their stakes and the policy issues they faced. The results of this research led Berman and Oettinger to outline a public policy agenda for addressing issues such as those raised by bills for a Consumer Communications Reform Act of 1976.

"Control over communications services is a great source of power in our society," writes Berman at the outset of the second study, *Computer or Communications? Allocation of Functions and the Role of the Federal Communications Commission.* With this source of power at stake, Berman discusses what he calls "compunications" —the marriage of the telephone and the computer—and the policy making headaches the combined technologies forced upon an FCC ill prepared to settle the resulting border warfare between the computer and communications industries. Borders set by the FCC in a 1971 decision had proved unacceptable by 1976, forcing the FCC—and the Congress—to take a deeper look at the whole matter.

The third study, by William H. Read, *Foreign Policy: The High and Low Politics of Telecommunications*, questions the adequacy of the international communications policy making process. Does the process remain satisfactory in light of the rise in information technologies which in turn have promoted the global condition known as interdependence? The premises for the inquiry are discussed in detail and a change is recommended by Read, who is a Program Fellow from the White House Office of Telecommunications Policy.

In reading these studies it will become apparent to the reader that, far from being definitive, these are initials cuts into subject matter that requires continued, thoughtful attention. The authors have not attempted to write the last word; to the contrary, their claim is to offer some introductory words for what they see as an emerging public policy debate on information resources.

High and Low Politics: Information Resources for the 80s

Part One

The Medium and the Telephone: The Politics of Information Resources

Paul J. Berman
Anthony G. Oettinger

1. INFORMATION RESOURCES IN PERSPECTIVE

Significance of the Information Resources Web

Information is essential to organization. The rule holds for the paramecium as for the Department of Defense. Information resources thus confer both power and vulnerability.

The eastern power blackout of 1965 and the Arab oil embargo of 1974 underscored our dependency on energy resources. The disruptiveness of even a partial loss of information services—inability to call the doctor, the broker, or the central computer—became clear to the residents and business people in the Gramercy area of lower Manhattan, whom a telephone exchange fire left without telephone service for several weeks early in 1975. The organizing role of information is manifest in the genetic code whereby DNA controls the life and propagation of the lowliest organisms; this organizing role grows with the complexity of individual organisms and social groups (Fig. 1–1) [1].

Our individual and collective welfare, in terms of both daily living and high politics, depends on the basic functions of creating, storing, processing, distributing, and using information—functions that pervade all social activity (Fig. 1–2). The industries that perform various combinations of these functions are also big business (Fig. 1–3). By the mid seventies, information occupations employed close to half the American labor force [2]. Domestically, pervasive information systems (Figs. 1–4, 1–5, 1–6) help shape both the personal elements of daily life—how we get billed, who keeps what kinds of personal records about us—and our major social systems—what news is gathered and spread, how well financial, transportation and other systems serve us, what military or economic intelligence is gathered for or about us, and so on.

Overseas calls to and from the United States rose from 8.1 million in 1965 to 60.0 million in 1974. Membership in Intelsat, the international communications satellite consortium, grew from 19 countries in 1965 to 75 in 1975 (Fig. 1–6). These growths reflect increasing information flows that enable transnational organizations to pursue global strategies, thereby promoting economic, social, and political interdependence and making control over telecommunications through government policies a powerful lever on international relations [3].

All our lives are thus greatly affected by decisions as to what information services are provided, who provides them, who gets them, how effectively, and at what prices. Such decisions emerge from the interplay of many individuals and organizations in the marketplace: from the laboratories and board rooms of private national and multinational corporations; from the many and lengthy proceedings of international bodies and national authorities such as the International Telecommunications Union, the Federal Communications Commission and the Federal Reserve Board; from the Congress, which considers bills such as the Consumer Communications Reform Act introduced early in 1976; from the courts, where a government antitrust suit against IBM has been underway since 1969 and another against AT&T emerged in 1974; from expert commissions, such as the National Commission on Electronic Fund Transfers, the National Commission on New Technological Uses of Copyrighted Works, and the Privacy Protection Study Commission, all created by Congress late in 1974; and, in many variations, from local and state authorities in each of the fifty states.

Traditionally, such diverse information systems as the print media, the telegraph, the telephone, the television networks and electronic data processing have been seen as providing distinct services for distinct users, through distinct technologies and market structures. Past public interventions reflected these perceptions. For example, although the Communications Act of 1934 appears to have established the Federal Communications Commission as a single body to regulate both broadcasting and interstate telephone enterprises, the Act has two distinct parts and the Commission two distinct bureaus that carry on the separate regulation of telephone service and broadcasting previously established by the Interstate Commerce Commission and by the Federal Radio Commission, respectively.

In 1975, most legislation about information systems antedated World War II. Yet the pervasiveness of information systems and our dependence on them had grown enormously in the three preceding decades. Also, information systems had grown so thoroughly intertwined (Fig. 1–2), that a tug on any one of them threatened to disturb all the others (Fig. 1–7). Yet the caricature of the legal mind as "one that could think of something that was inextricably connected to something else without thinking about what it was connected to" aptly describes conventional thought and action about information systems [4]. The resulting strains on how we manage information resources are not spotlighted by an "information crisis" like the "energy crisis," but they are nonetheless severe.

Tracing Through the Web

Our aim here is to sketch the web of relations among ongoing and prospective major changes in information systems, in order to illuminate the strains and to contribute toward widespread understanding by sovereign citizens of the processes for deciding what information services are provided, who provides them, how, to whom, and at what prices. We shall portray, so far as we can, the fulness of the complex interactions among the diverse political, economic, and technological factors that shape the "infrastructure" of information systems. By so doing, we aim to give a broad perspective which, because of necessary and useful specialization, is generally absent from advocacy by special interests and from in-depth studies by academics or professionals. In conclusion, we suggest that there are issues more subtle and policy choices more critical than is made evident by those who tout information systems as technological fixes for this or that social ill, those who believe that technological change is evil and inexorable, or those who think that their institutions have manifest and sacred destinies.

Tracing through the whole web of information systems is still far beyond our intellectual means. We limit ourselves here to the domestic scene in the United States. Worldwide developments, though closely related to the domestic, are treated elsewhere [5]. We also need important threads to follow, rich in significant cross-links to other threads. Computer and telecommunications technologies form one such thread. They have merged into *compunications*, a common, overarching set of digital information technologies. As shown in Figure 1–2, compunications systems supply most of the infrastructure not only of all information industries, but of most other activities as well. Vast and fast scientific and technological advances have taken place in compunications since World War II (Figs. 1–8, 1–9, and 1–10) and promise to continue well into the next decades. The strains on traditional thinking and policy making emerge particularly plainly from analyzing this realm, and the need for fresh and wide understanding and informed policy making is correspondingly great.

Among the many uses of compunications systems, we single out as our main thread the gathering and spreading of news. Figure 1–3 shows news media to be embedded in industries of relatively modest, yet significant financial size. Their products are familiar and important to nearly everyone; that importance, as reflected in the First Amendment to the Constitution of the United States, transcends mere financial or technical measures. Since the news has a recognized and especially high leverage on human affairs, it aptly illustrates how changes in the nature of information products can affect the nature of everything else in a society. We trace how changes in other functions of compunications systems can affect the news advertently or inadvertently. This is our main source of concrete and specific examples of the complex interactions among the important facets of information systems.

One institution will unavoidably dominate any discussion of information in America: the American Telephone and Telegraph Company and its subsidiaries—the Bell System—which, along with all the smaller independent telephone companies, has won a place in every information transfer scheme worthy of the name. Nearly all the connecting lines in Figure 1–2 (which shows the merging of computation and communications services) are telephone company lines.

By way of background, we begin, in Section 2, by sketching selected technological traits of the network of telecommunications facilities. We list the vital functions that this network performs for all the news media besides carrying their conversations. Section 3 describes how radio and television news came to depend on common carrier facilities and services; it shows how combined economic, political, and technological factors led to choosing this network as the backbone of the U.S. broadcasting system from among the many possibilities available at radio's birth.

The news media, like everybody else, use ordinary telephones heavily for ordinary conversations, through what the telephone companies call local service and message telecommunications service (long distance). They are also among the important users of the so-called private line services. The role of private line services is sketched in Section 4. There, a case is made for the proposition that the prosaic question of how much the news media are charged for private line services is at least as important in determining the source, content, and dissemination of the news as the sinister factors so often alleged in rhetoric from all across the political spectrum.

As Figures 1–2 and 1–7 suggest, it is not the news alone that could be significantly affected by the prices of private line services. Nor, as it happens, are private line prices independent of all other telecommunications prices. Indeed, by setting the terms of access—both absolute and relative—it is the web of telecommunications pricing that binds together all telecommunications users in the world. The news is just a convenient example and an interesting one. The news media are not among the largest spenders on telecommunications. Telephone conversations, airline navigation or reservation services, brokerage services, electronic funds transfers, the U.S. government's Federal Telecommunications System, and many many others could have served as examples, since all can be affected by changes in telecommunications pricing policy, whatever the causes of change might be.

Section 5 goes to the heart of the matter with a detailed look at telecommunications pricing processes. We show how ambiguously economic theory accounts for actual pricing processes and how political and technological factors are essential toward understanding them. In particular, we show that the total costs of all services are roughly equated to the total revenues of all services, but that, within this overall equation, there is much latitude in the pricing or particular services such as the private line services. So long as the overall equation must hold, revenues lost in one service must be gained in others or vice versa; hence the strong linkages among all users of all services, including the news media and every ordinary citizen.

The processes outlined in Section 5 grew out of many skirmishes in the century since Alexander Graham Bell invented the telephone. Since the sixties, these processes have come under renewed and intense challenge. The small-consumer interests traditionally concerned about the rate increases of public utilities were increasingly more active and better organized in keeping with the times. But they have played only insignificant direct roles in this arena. The most persistent players include local, state, and federal regulators and the telephone companies themselves. Significantly, however, the growth of the electronics industry in general and of the computer industry in particular has introduced new potentials for technological and price competition with traditional telecommunications institutions, especially in the private line services. New or at least increased service demands by major corporate consumers and their search for lower prices have also spurred interest and entry in new and old markets.

The fresh combatants and the altered battlefields are sketched in Section 6.

What the new situation means to the news media is shown in Section 7. In brief, the media were alarmed that they might fall victim to crossfire between the old regime and the young pretenders. All the organizations listed in Figure 1–7, and many others, have voiced similar concerns. Thus, however remote the telecommunications pricing battles might seem from workaday problems, they are widely perceived as capable of strong impacts on all organizations and ordinary citizens. Section 8 summarizes diverse perceptions of the issues and the stakes in all the battles described in the preceding sections. In conclusion, Section 9 suggests the following questions for a policy agenda to address these perceived issues:

1. What should be our policy aims for the 1980s and beyond?
2. What is to be the scope of policy for the 1980s and beyond?
3. Who is to decide what the aims and scope of policy should be?
4. What are the limitations of present policy instruments?
5. What policy instruments—institutions and processes—might best serve our aims for the 1980s and beyond?
6. How best to match up industry structure with performance aims?
7. How best to harmonize domestic and foreign policy instruments?

(Text continues on page 29 . . .)

TYPE OF TRANSACTION	AVERAGE ANNUAL GROWTH 1940–1970
Checks written	6.7%
Telephones in use	6.2
Individual Social Security payments	17.1
Individual federal tax returns	5.8
Public welfare recipients	3.5
Airline passengers	14.3
Persons entering hospitals for treatment	3.8
Persons covered by private hospitalization insurance	9.5
Motor vehicle registrations	4.0
Passports issued	16.0
Students enrolled in colleges and universities	5.2
Applications received for federal employment	3.7
New York Stock Exchange transactions	8.3
Pieces of mail handled, U.S. Postal Service	3.6
Abstracts produced by Chemical Abstracts Service	6.5
Local telephone conversations	6.5
Long-Distance telephone conversations	9.5
U.S. population	1.5
U.S. gross national product	3.9

Figure 1–1. Growth of Transactions and of Complexity

The number of transactions of various types, and of information records associated with these, has increased much faster than the U.S. population over the last three decades. This rapid growth of interactions among individuals and institutions is one rough index of the growing complexity of society.

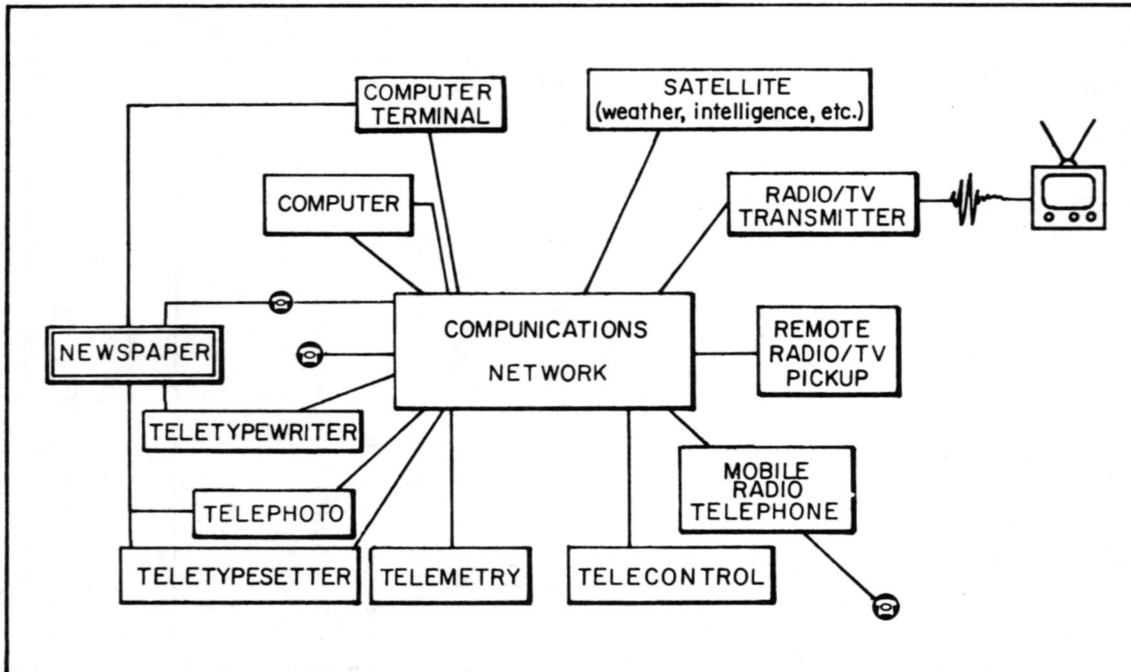

Figure 1–2. Social Infrastructure: The Changing Telecommunications Network

As of 1974, telephones still predominated—144 million in the U.S., 40 percent of them extensions—but many other devices are now attached to a network that has become an infrastructure basic to most social functions including many that reach directly into the home. As computers and computer terminals have become increasingly pervasive over the last two decades, the network has developed toward an integrated computer/communications, or *compunications*, network.

(in billions of dollars;
double counting not eliminated)

	1970	1971	1972	1973
Broadcast television	2.8	2.8	3.2	3.5
Cable television	0.3	0.3	0.4	0.5
Broadcast radio	1.1	1.3	1.4	1.5
Telephone	18.2	20.0	22.4	25.5
Telegraph	0.4	0.4	0.4	0.5
Specialized common carriers	0.0	0.0	0.0	0.0
Satellite carriers	0.1	0.1	0.1	0.1
Mobile radio systems	2.0	2.2	2.4	2.6
Motion pictures	3.8	3.8	NA	NA
Organized sports, theaters, etc.	4.4	NA	NA	NA
Computer software suppliers Computer service suppliers	1.9+	2.4+	3.0+	3.7+
U.S. Postal Service	6.3	6.7	7.9	8.3
Private information delivery services	0.7+	0.8+	1.0+	1.2+
Newspapers; wire services	7.0	7.4	7.8	8.3
Periodicals (including newsletters)	3.2	3.4	3.5	3.7
Business consulting services	0.9	1.1	NA	NA
Advertising	7.9	7.6	NA	NA
Marketing	32.4	37.7	41.3	43.4
Brokerage industries	40.6	47.4	54.4	NA
Book publishing and printing	3.4	3.7	3.9	4.1
Libraries	2.1	NA	3.6	NA
Schooling	70.0	76.3	83.2	89.5
Research and development	26.5	27.3	29.2	30.6
Federal information institutions				
Census Bureau	0.1	0.1	0.1	0.1
National intelligence community	4.0+	NA	NA	NA
National Technical Information Service	0.0	0.0	0.0	0.0
Social Security Administration	1.0	1.2	1.3	1.4
County agents	0.3	0.4	0.4	0.5
Banking and credit	61.1	68.9	76.9	NA
Insurance	92.6	103.5	121.4	NA
Legal services	8.5	9.6	NA	NA
U.S. Gross National Product	**977.1**	**1,055.5**	**1,155.2**	**1,294.9**

Figure 1–3. The Information Industries

The information industries are those where creating, storing, processing, distributing, or using information are *primary* functions. Elsewhere, these functions are *incidental* to some other primary role. Even the massive information processing performed by railroads or supermarkets qualifies them as important information users, but not as information industries. Approximate annual gross revenues shown here for 1970–73 provide a rough index of the relative sizes of these industries. Double-counting has not been eliminated.

RULES FOR USING THE TELEPHONE—CIRCA 1909

AN AMUSING look at some of the by-laws of the early days of the Walker County Telephone Company is provided in these 1909 by-laws of the telephone company as passed by the Board of Directors:

(1) Each person renting a phone shall accept the same subject to the by-laws of the company and such regulations as may be fixed by the company from time to time.

(2) Social conversations are not permitted when the line is required for the transmission of business messages.

(3) No one shall use the line for more than five minutes if another is waiting for it. Anyone failing to surrender the line in five minutes after being requested to do so shall be charged the sum of 25¢ for each three minutes and at the same rate for any extra time over.

(4) Any and all persons using party lines shall, upon notice that there is a long distance message to be transmitted, immediately surrender the line, and in case of failure to so surrender, each of said parties using the party line shall pay 25¢ for each three minutes and at the same rate for any time over three minutes.

(5) No one shall be allowed to use improper language over the line and any person so offending shall be charged $1.00 for such conversation over the phone and will also be liable for prosecution for violation of the law.

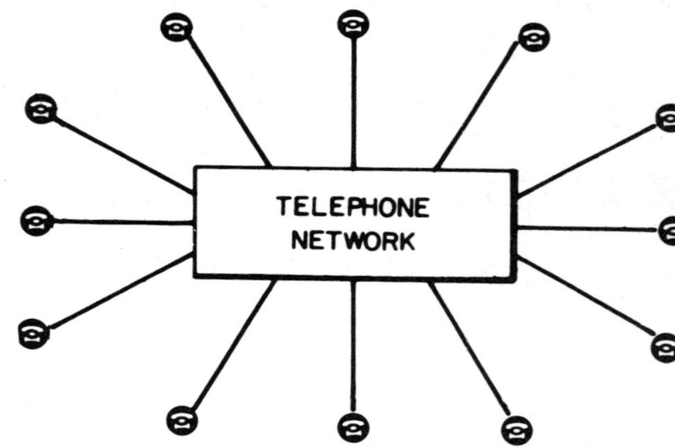

(6) No one shall be allowed to take down the receiver for the purpose of listening to a message not intended for him. Each person so offending shall be charged 25¢.

(7) No one shall allow anyone but members of his family, invited guests or other subscribers to use his telephone free of charge. All subscribers allowing messages from their phone by parties not authorized to talk free will be charged with the conversation (10 cents) and shall be required to collect the same and turn it over to the telephone company at the end of each month.

Figure 1–4. Changing Perceptions of the Telephone

The differences between personal and business uses of the 150 million telephones of today and the few at the start of the century are evident from the contrast between these 1909 rules and our own present practices. The contrast is further sharpened in Figure 1–2, which showed how many information services, other than conversations over ordinary telephones, increasingly depend on the infrastructure provided by the telecommunications network.

Source: *Communications News*, Vol. 12, No. 1, January 1975, p. 24.

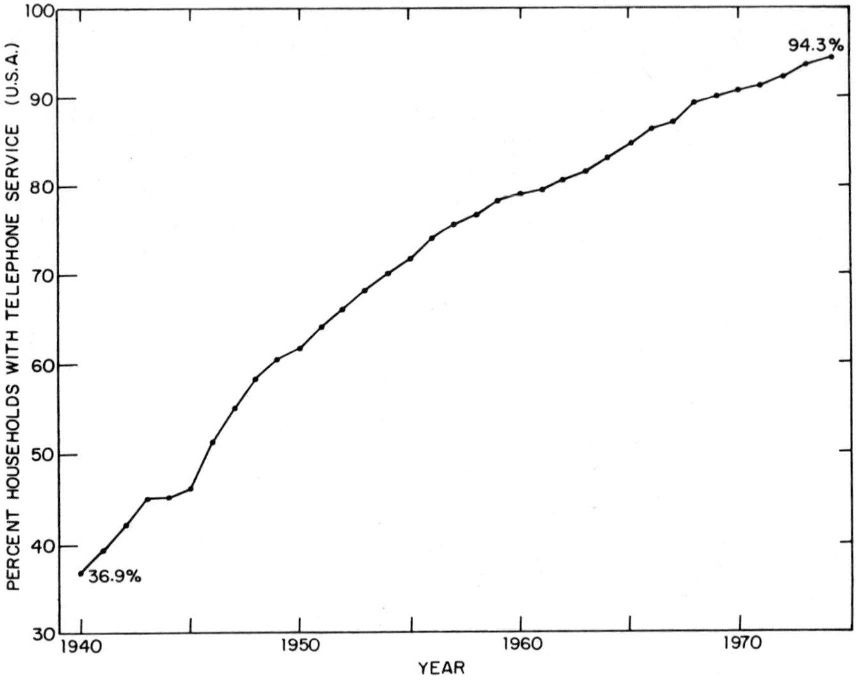

(a)

Figure 1–5. Diffusion of Telephone and Computer Usage

The increasing pervasiveness of the telephone is illustrated by the growth of U.S. households with telephones from 37 to 94 percent in 1974 (a). Related statistics are shown for comparison (b).

Source: U.S. Population: U.S. Department of Commerce, Bureau of the Census, *Statistical Abstract of the United States, 1974*, 95th edition (Washington, D.C.), 1974, Table No. 2, p. 5.

U.S. GNP: *Ibid*, Table No. 598, p. 373.

Bell System Operating Revenues: American Telephone and Telegraph Company, *Bell System Statistical Manual*, §100—Earnings, p. 102.

Number of Telephones: *Ibid*, §500—Telephones, p. 504.

Percent of Households: *Ibid*.

Figure 1–5. cont.

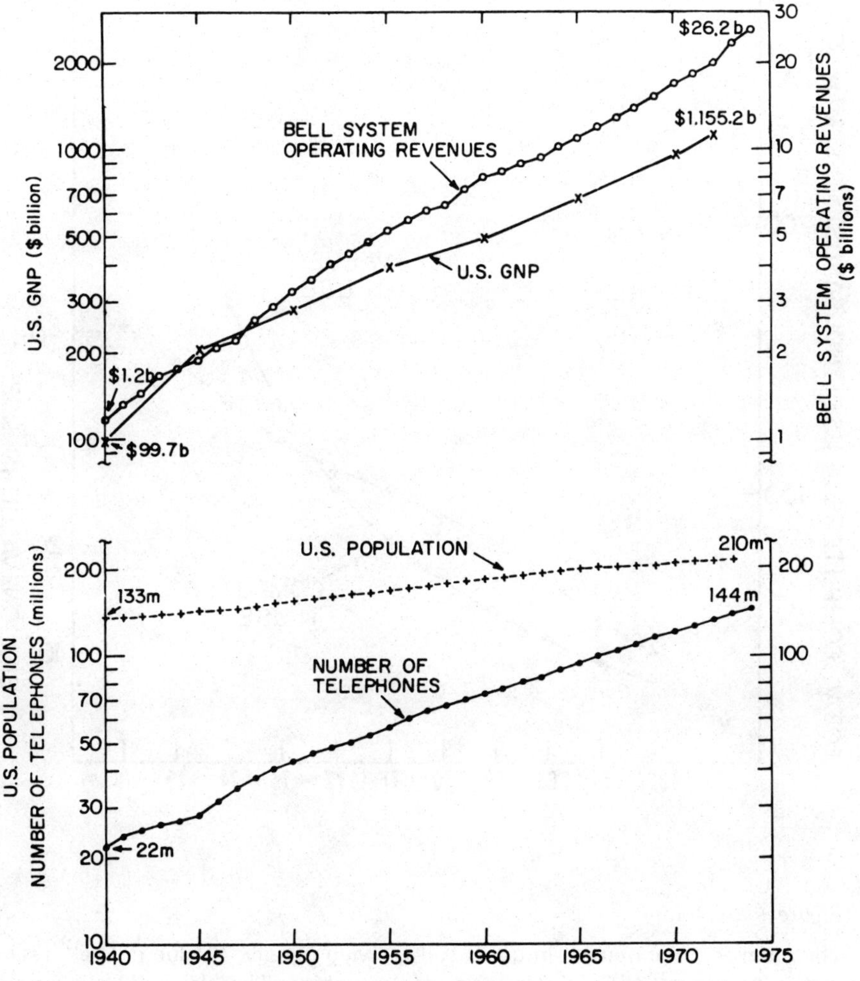

(b)

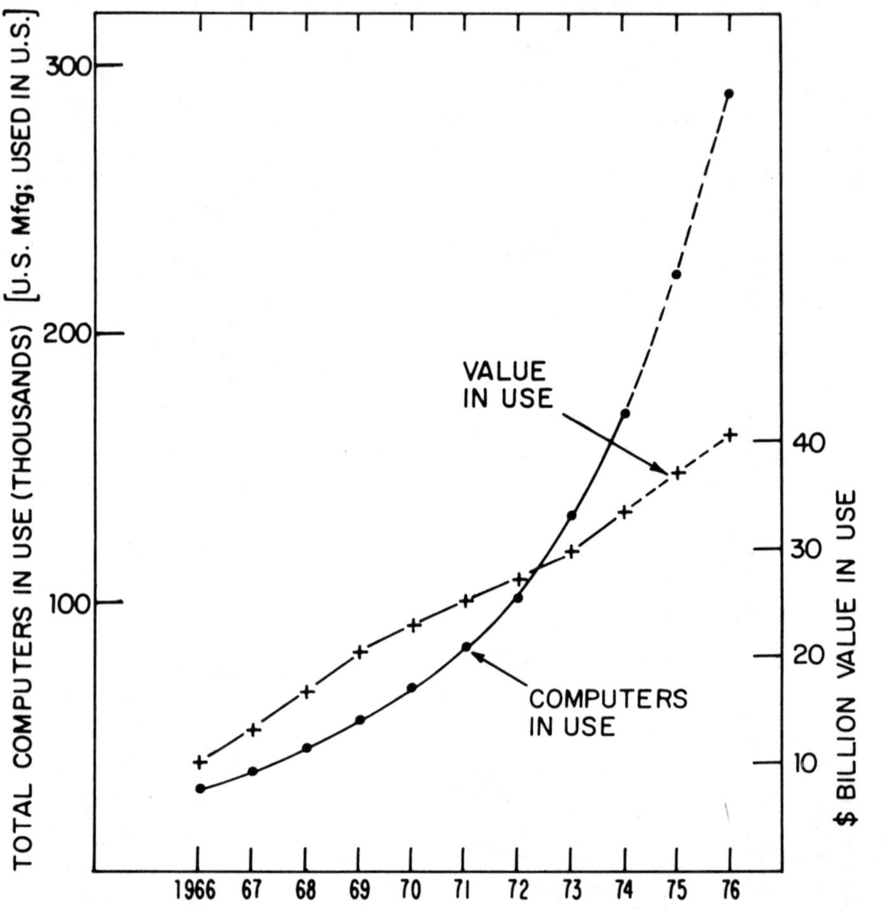

(c)

Figure 1–5. cont.

The number of computers in use is still growing rapidly. "Value in Use" is estimated by accumulation of sales price of units shipped less value of units retired from service. The introduction of the $20–100 pocket electronic calculator since 1972 has triggered a significant transition from computers as solely industrial goods toward computers as consumer goods as well. These calculators are not counted in the graph.

Source: John Breyer, and James Peacock. International Data Corporation, *Testimony and background material before U.S. Senate Committee on the Judiciary, Subcommittee on Antitrust and Monopoly*, July 23, 1974, Table 112, p. 27.

	State	Percent of Households with Telephones		State	Percent of Households with Telephones
1	New Hampshire	102	26	Missouri	94
2	New Jersey	102		United States	94
3	Connecticut	101			
4	Delaware	101	27	Virginia	94
5	Hawaii	100	28	Florida	93
6	Vermont	100	29	Nevada	93
7	Rhode Island	99	30	Washington	93
8	Illinois	98	31	Wyoming	93
9	Maine	98	32	Indiana	91
10	Massachusetts	98	33	Oklahoma	91
11	Michigan	98	34	Texas	91
12	Minnesota	98	35	Arizona	90
13	Nebraska	98	36	Georgia	90
			37	Idaho	90
	District of Columbia	97	38	Louisiana	90
			39	Montana	90
14	Maryland	97	40	North Carolina	90
15	North Dakota	97	41	Oregon	90
16	Pennsylvania	97	42	Tennessee	89
17	Utah	97	43	Kentucky	87
18	California	96	44	South Carolina	87
19	Iowa	96	45	West Virginia	86
20	Ohio	96	46	Alabama	85
21	South Dakota	96	47	Arkansas	84
22	Wisconsin	96	48	New Mexico	84
23	New York	95	49	Mississippi	79
24	Colorado	94	50	Alaska	77
25	Kansas	94			

(a)

Figure 1–6. Distribution of Telephones in the United States and the World

Although impressively pervasive in the aggregate, telephones and computers are not uniformly distributed, in either the U.S. or the world.

Alaska's 77% of households with telephones as of the beginning of 1975 contrasts with 102% in New Hampshire and New Jersey. Percentages greater than 100 arise from dividing the number of telephones by the number of households (e.g. families), rather than by the number of homes. Thus, a state inhabited by one family with two homes, each with a telephone, would show 200% of households with telephones, but 100% of homes.

Source: American Telephone and Telegraph Company, Comptrollers Accounting Division, April 1975, "Telephone Development by States, January 1, 1975."

Rank Ordering of Various Countries

EDP base as % of GNP	Phones per 100 pop.	GNP per capita	Country	GNP per capita (in dollars)
4	3	1	Switzerland	5,933
1	1	2	United States	5,526
13	2	3	Sweden	5,463
8	11	4	West Germany	4,710
5	4	5	Canada	4,682
3	5	6	Denmark	4,540
7	15	7	France	4,250
10	6	8	Australia	4,054
14	7	9	Norway	4,050
15	12	10	Belgium	3,880
9	10	11	Netherlands	3,846
16	13	12	Austria	3,257
6	8	13	Japan	3,187
2	9	14	United Kingdom	2,714
20	18	15	USSR	2,223
12	14	16	Italy	2,185
18	16	17	Spain	1,491
			Worldwide Total	1,147
19	19	18	Yugoslavia	1,129
11	17	19	South Africa	830
21	20	20	Mexico	747
17	21	21	Brazil	503

(b)

Figure 1–6. cont.

The rank ordering of countries by GNP per capita is strongly correlated with rank ordering by both computers as percent of GNP and telephones per 100 population, although there are also some interesting anomalies.

Source: International Data Corporation, *Testimony and Background Material before the U.S. Senate Committee on the Judiciary, Subcommittee on Antitrust and Monopoly, July 23, 1974* (Newtonville, Mass.: IDC), July 1974, Table No. 135, p. 81.

United Nations, *Statistical Yearbook, 1973* (New York, N.Y.: U.N.), 1974, pp. 495–498.

	Intelsat	
	1965	1975
Member countries	19	75
Voice/grade circuits (4 kilohertz) in use	75	10,050
Earth stations	5	73
Total investment in space segment (includes launches, everything in orbit and their maintenance)	$8,000,000	$450,000,000
Annual unit cost per circuit (based on full lifetime and full capacity satellite)	$32,000	$9,000

(c)

Figure 1–6. cont.

International satellite communication links have grown rpaidly in the last decade, alongside undersea cables.

Source: Neil R. Helm, COMSAT Laboratories, Handout, Institute for Graphic Communications Conference on World Telecommunications, June 8, 9 and 10, 1975.

Growth in World's Telephones *(In Millions)*

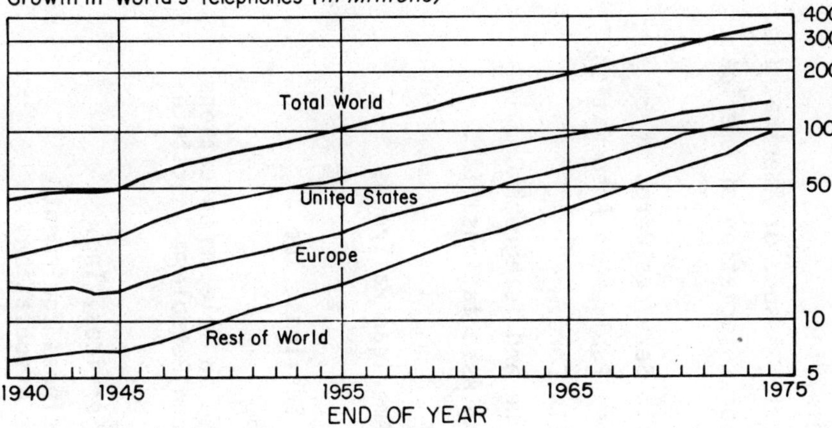

(d)

Figure 1–6. cont.

Although the U.S. alone still has more telephones than all the rest of the world combined, the growth rate outside the United States is high.

Source: American Telephone and Telegraph Company, *Bell System Statistical Manual*, §500—Telephones, p. 503.

Aeronautical Radio, Inc.
Aerospace Industries Assoc. of America, Inc.
Air Transportation Association of America
Altair Airlines
American Facsimile Services Corp.
American Newspaper Publishers Association
American Satellite Corp.
American Telephone and Telegraph Co.
American Trucking Associations, Inc.
Associated Press
Association of American Railroads
Bank Wire
Boeing Computer Services, Inc.
Bunker Ramo Corp.
Central Committee on Telecommunications of the American Petroleum Institute
Citicorp
Commodity News Services, Inc.
Computer and Business Equipment Manufacturers Association
National Association of Manufacturers
National Association of Motor Bus Owners
National Retail Merchants Association
National Data Corp.
North American Telephone Assoc.
Orlando Communications Club, Inc.
Packet Communications Inc.
RCA Global Communications, Inc.
Remote Data Processing Services Section of Adapso, Inc.
Satellite Business Systems, Inc. (formerly CML Satellite Corp.)
Securities Industry Automation Corp.
Southern Pacific Communications Company
Telenet Communications Corp.
Tymshare, Inc.
United Press International, Inc.
United System Service, Inc.
Utilities Telecommunications Council
Wells National Services Corp.

Data Transmission Company	Western Union International, Inc.
Dow Jones & Company, Inc.	Western Union Telegraph Company
Graphnet Systems, Inc.	Xero-Fax, Inc.
GTE Service Corp.	
International Business Machines	Department of Health, Education and Welfare
ITT World Communications Inc.	Department of Justice
MCI Telecommunications Corp.	Office of Telecommunications Policy

Figure 1–7. Interdependence of Information Systems

Illustrative of the multiple stakes in the structure and pricing of basic information services is the density of the organizations commenting on an inquiry into resale and shared use of telecommunications services and facilities that the Federal Communications Commission originated in mid 1974 (Docket No. 20097). Some service companies or trade and other associations have found a niche in purchasing wholesale and in retailing (sometimes with other services added) on terms more favorable than those available directly from the common carriers. Hitherto, such brokerage has not been common in telecommunications; its effect on the structure of the telecommunications industry, questions about boundary between telecommunications and data processing, and allegations of illegal preferential treatment are among the issues raised. In part, the inquiry arose when truckers (American Trucking Associations, Inc.) alleged preferential treatment of airlines (Aeronautical Radio, Inc.).

Also at stake are interests of newspapers and news wire services, banks, manufacturers and retail merchants; of various suppliers of telecommunications and data processing goods and services; and of the general "public interest, convenience, and necessity." Television interests are not directly represented here. As users of lines of much higher capacity than those at issue in Docket 20097, they participate in more directly relevant proceedings. This fragmentation of proceedings masks strong linkages that are brought more clearly into view in Section 5 and following sections.

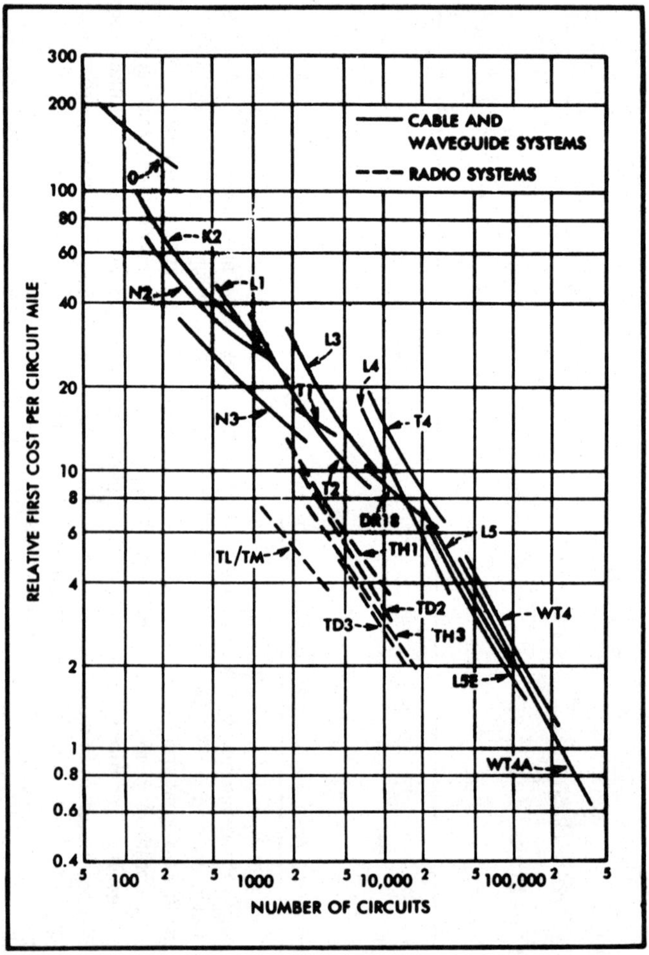
300
200
100
80
60
40
20
10
8
6
4
2
1
0.8
0.6
0.4
RELATIVE FIRST COST PER CIRCUIT MILE
CABLE AND
WAVEGUIDE SYSTEMS
RADIO SYSTEMS
O
K2
N2
L1
L3
L4
N3
T1
T4
T2
DR18
L5
TL/TM
TH1
TD2
TD3
TH3
WT4
L5E
WT4A
5
100
2
5
1000
2
5
10,000
2
5
100,000
2
5
NUMBER OF CIRCUITS

(a)

LEGEND		
Bell System Code	Transmission Vehicle	Range
DR	Digital Microwave Radio	Short
K	Cabled Wires	Long
L	Coaxial Cables	Long
N	Cabled Wires	Short
0	Open Wires	Short
T1	Digital Cabled Wires	Short
T2	Digital Cabled Wires	Long
T4	Digital Coaxial Cables	Long
TD,TH	Microwave Radio	Long
WT	Waveguide	Long

Figure 1–8. Telephone Cost and Price Trends

Cost trends for short and long haul transmission are down while carrying capacity is up (a), both by several orders of magnitude. Each of the vehicles illustrated here is capable of carrying simultaneously many conversations or other types of information. This includes the "0" system, which improves on the classical open wires. The increase in the handling capacity of the several vehicles represent developments over time as well as technical improvements. Line termination, switching, and personnel costs are among the additional elements that enter into the cost of complete telecommunication systems. Present cost proportions and price patterns are such that a pair of wires carrying but one conversation is still generally the vehicle for local conveyance from an individual telephone toward the nearest exchange, where many conversations heading the same way are bundled for further short or long haul.

Source: Eugene F. O'Neill, "Radio and Long-Haul Transmission," *Bell Laboratories Record*, January 1975, p. 55.

Figure 1–8. cont.

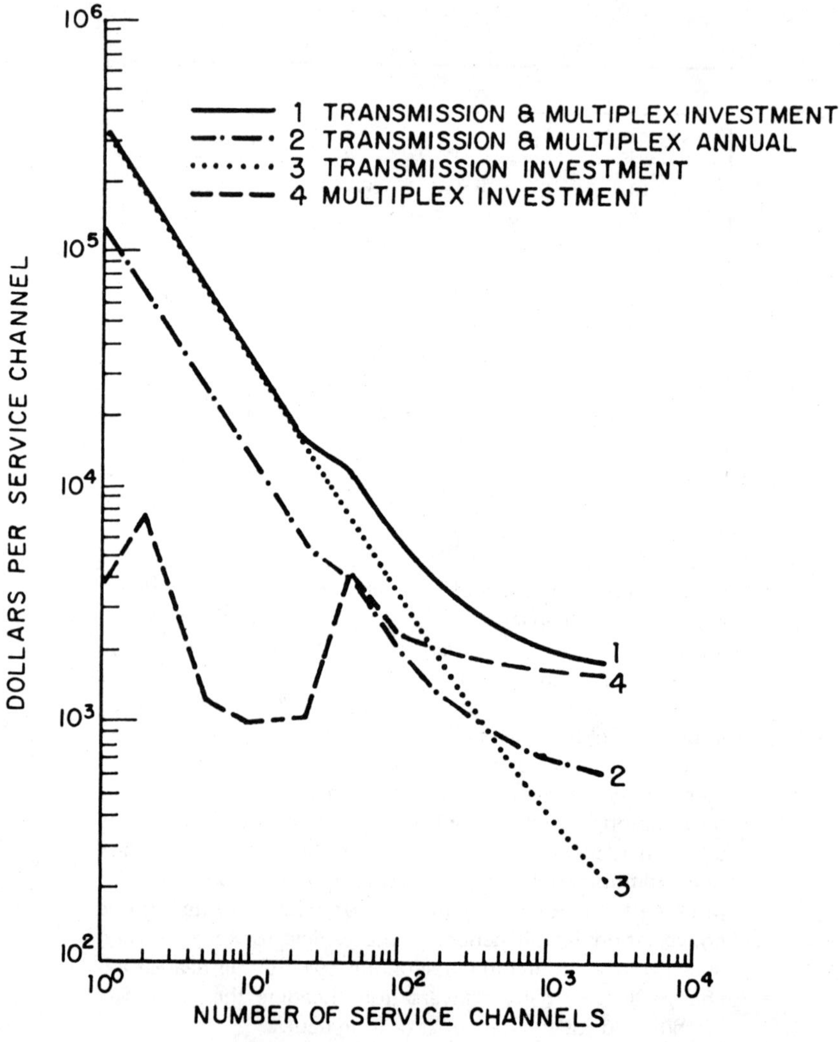

HIGH CAPACITY DIGITALLY AUGMENTED MICROWAVE

500 MILES

(b)

Figure 1–8 (b). cont.

So-called economies of scale are reflected by the downward trend in (a), where transmission costs per circuit decrease as the capacity of the system increases. At full load, for example, the WT4A system is both larger and more economical than its predecessors. The precise nature of economies of scale is hotly disputed, since their absence favors the effectiveness of competition, while their strong presence defines a "natural" monopoly. While the cost of transmission alone decreases smoothly with scale, as shown in (a) and in (b/3), multiplexing costs introduce lumps, as shown in (b/1,2). For zero transmission costs, the lumpiness would dominate (b/4). Multiplexing is the technique used to carve multiple channels out of a single facility.

The picture is further complicated by distinctions between comparing technologies over the long term, assuming a fixed loading, and the short term effects on unit costs of the transition between no load and full load of a fixed facility. What one party sees as prudent spare capacity may be seen as wanton overcapacity by another.

Source: George P. Mandanis et al., *The Domestic Telecommunications Industry: Economic Behavior, Competition, and Public Policy*, R73–40 (San Rafael, Calif.: Systems Applications, Inc.), May 10, 1974, p. 33.

Bangor	-	Miami	1.05
Bangor	-	Portland	0.89
Portland	-	New York	0.88
New York	-	Washington	0.85
New York	-	Richmond	0.80
New York	-	Miami	1.12

(c)

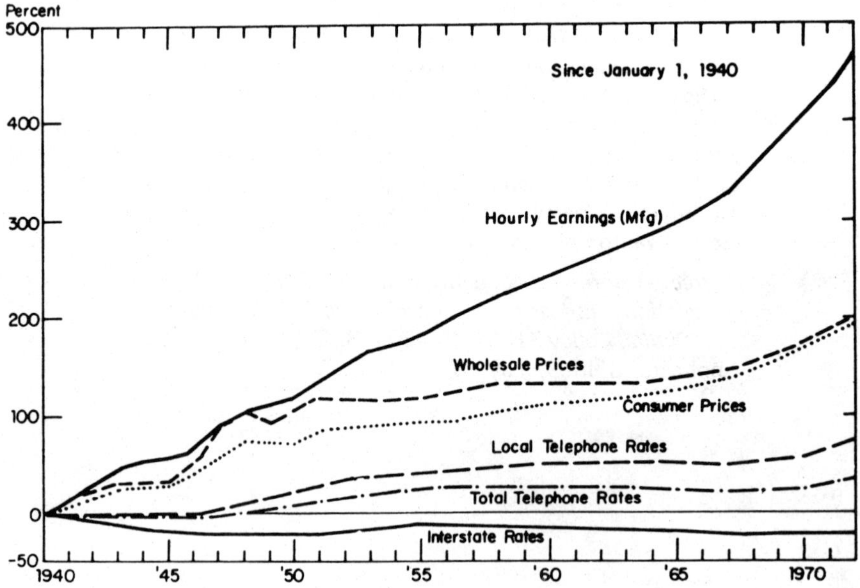

(d)

Figure 1–8 (c). cont.

The inventory costs of carrying unused "lumpy" plant while it fills up affect average costs over the period of averaging. Depending on the size of the lumps and assumptions about demand (rate of fill), "better" technology may or may not realize economies of scale. Under a fixed set of assumptions about startup demand, rate of demand growth, traffic density (cross-section) along each route, length of route, subscriber density in local service area, etc. new equipment may or may not reduce costs. For six private line voice routes studied by Systems Applications, Inc., for the White House Office of Telecommunications Policy, the projected ratios of average costs in 1980 to average costs in 1972 were less than 1 (economies of scale realized) in four cases, greater than 1 in two cases.

Source: George P. Mandanis et al., p. 231.

Figure 1–8 (d). cont.

Downward transmission cost trends are reflected in part in total telephone prices rising less rapidly than overall price indexes. The rise in local rates and drop in interstate rates are accounted for by the interplay of technological, economic, and political factors.

Source: U.S. Congress, Senate, Committee on the Judiciary, Subcommittee on Antitrust and Monopoly, Hearings on S.1167, Part 2: The Communications Industry, July 30 and 31; August 1 and 2. Data submitted by American Telephone and Telegraph Company, p. 1721.

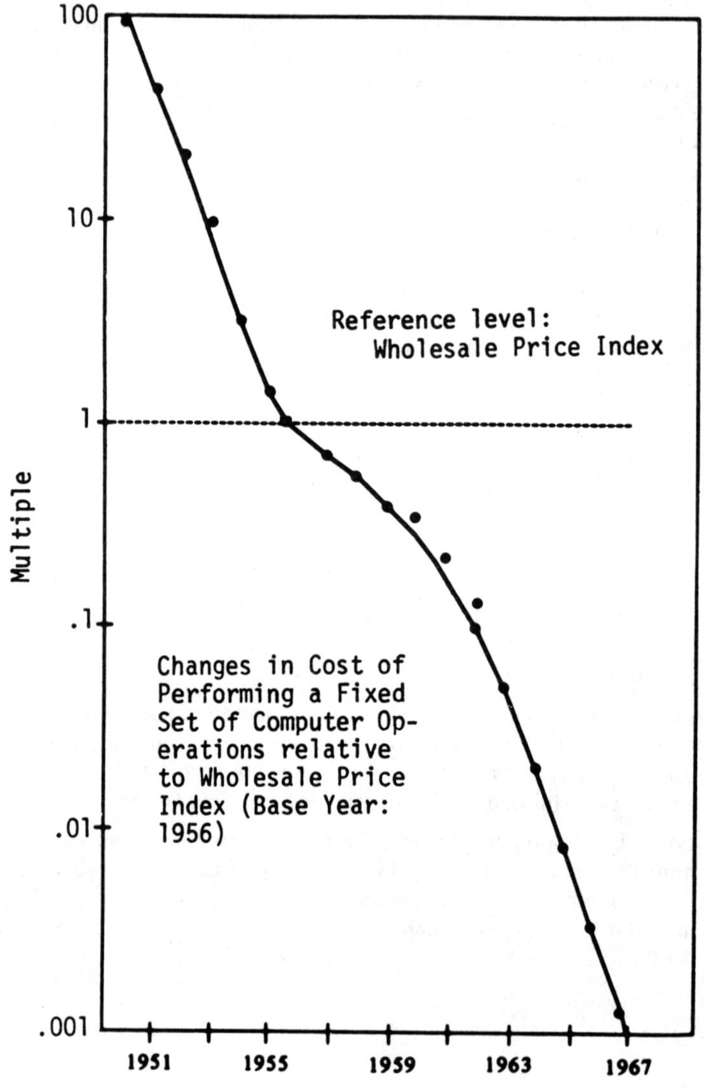
100
10
1
.1
.01
.001
Multiple
Reference level:
Wholesale Price Index
Changes in Cost of
Performing a Fixed
Set of Computer Op-
erations relative
to Wholesale Price
Index (Base Year:
1956)
1951
1955
1959
1963
1967

(a)

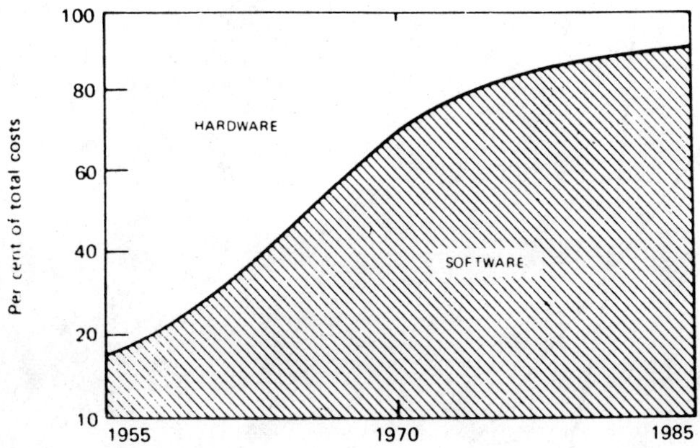

(b)

Figure 1–9. Computer Cost Trends

(a) Like transmission costs, *basic* computer costs show remarkable declines relative to general price indexes. Capacity increases are also dramatic. Changes in overall systems cost and capacity are not necessarily that dramatic, as suggested by (b).

(b) The dangers of translating cost decreases in components or in elementary operations into cost decreases in useful systems are illustrated by the rising proportion of software costs that limit the effects of dropping hardware costs.

Actual prices to consumers are further determined by economic and political factors discussed throughout the text.

Sources: (a) William Baumol and Mitityahu Marcus, *Economics of Academic Libraries* (Washington, D.C.: American Council on Education), 1973, p. 48.

(b) Barry Boehm, *Software and Its Impact: A Quantitative Assessment* (Santa Monica, Calif.: RAND Corp.), December 1972, p. 5.

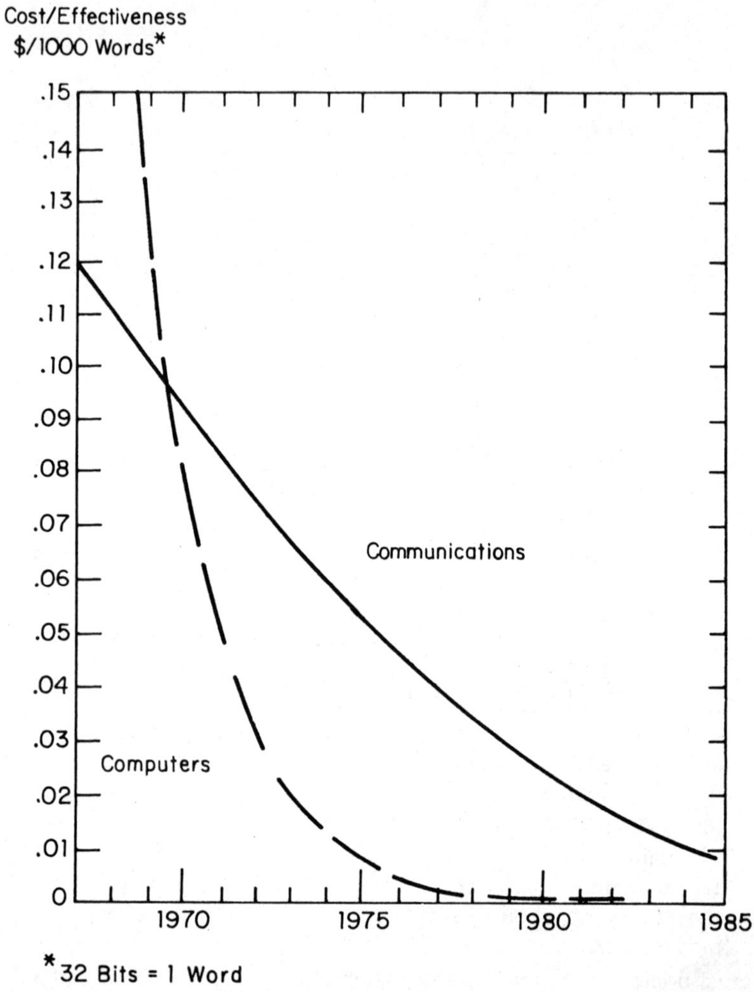

Figure 1–10. Comparative Computer Telecommunications Costs

It is widely believed that computer costs are dropping more rapidly than telecommunications costs. This is illustrated here for a system with relatively fixed computer and telecommunications requirements. How much such comparisons depend on technology and how much on pricing and market factors is not clear. At stake is how much data processing will be done through computer/telecommunications networks and how much through independent computer systems. (See Montgomery Phister, Jr., *Shared Systems versus Stand-Alone Systems*, Cambridge, Mass.: Program on Information Technologies and Public Policy, Harvard University, January 1975.)

Source: The Diebold Group, Inc.

2. TELECOMMUNICATIONS SYSTEMS—CONCEPTS AND REALITIES

The classic model of a telecommunications channel, as described by Claude E. Shannon in 1948 (Fig. 1–11), shows one transmitter and one receiver. As of 1974, there were some 360 million telephones in the world, 144 million of them in the United States. Any customer could therefore talk to about 10^8 others. Figures 1–12 and 1–13 show something of what that scale entails, given that the number of possible pairwise links is over 10^{17}. The real telephone network does not give each customer a permanent link to every other; it relies instead on switching in a hierarchical design based on bundling or "trunking" calls among exchange offices and higher centers (Fig. 1–13).

Since the effectiveness of switches and trunks depends on how many links might be active at any given time, the statistical distributions of the number of calls and of call durations (holding times) are important design parameters that depend on whether the links are used for conversations among people, for interactions among people and computers, or for other purposes. Other important parameters include channel bandwidth (in hertz), a measure of the range of pure tones that the channel can faithfully transmit. A voice telephone channel typically has a bandwidth of 3,000 hertz.

Alternatively, some measure of digital or pulsed channel capacity is used. Pulse transmission capacity is described in *bauds* (signal elements—or pulses of various heights—per second). How much information (as measured in *bits*) can actually be transmitted over a channel also depends on how much noise—hum, error, crackle, or snow—there is in the channel, as measured by the signal-to-noise ratio. For example, if the signal-to-noise ratio permits distinguishing 4 levels of pulse height, a channel may carry 2 bits per pulse, i.e., 2 bits per second per baud. On the other hand, if fighting noise requires sending redundant pulses, as through the use of error detecting codes, the bit rate may be less than the number of baud. Present technology affords roughly 3 bits per second per hertz of voice-grade bandwidth.

Especially significant for computer communications is access delay, or the time it takes to make a connection between transmitter and receiver. Network flexibility and growth potential, widespread availability of services and security that helps to protect confidentiality, are other performance traits of various degrees of importance. Continuity of service is a trait increasingly in the public consciousness, since the eastern power blackout of 1965, the Manhattan telephone fire of 1975, and sabotage and airliner hijackings have exposed the vulnerability of many essential services (Fig. 1–14).

The concrete facilities embodying abstract network functions are not evident in Figure 1–13, but figure 1–15 gives a snapshot of these facilities in terms of the capital invested in them; and it illustrates the dynamics of recent changes in the switching and transmission components. Figure 1–13 oversimplifies in other critical ways. Ordinary telephones, although predominant, are not the only instruments at the ends of telephone lines. As early as 1878, a prospectus for the District Telephone and Automatic Signal Company offered to connect burglar alarms to that company's central office via telephone wire, but the service apparently was unsuccessful [6]. Figure 1–2, above, showed devices currently linked by the telecommunications network. Telemetry and telecontrol devices are of obvious importance in the production of most goods as well as in science and technology. By linking remote pickups to radio or TV stations, and the stations with one another and with their transmitters, the telecommunications system provides the backbone of radio and TV networks. Teletypewriters link up the telecommunications network with the Postal Service as in the delivery of Mailgrams; along with telephoto terminals, teletypewriters have long been the workhorses of news services such as the Associated Press (AP) and United Press International (UPI), and of most newspapers and radio/TV news services as well. Computers, computer terminals, and telecommunications facilities have recently been integrated into a compunications system that automates many of UPI's news collecting, editing and distributing functions. In an increasing number of newspapers, particularly middle-sized chains, they are the primary tools of editors and of advertising copy writers and increasingly they are used both by reporters and in the production process.

This sharing of elaborate facilities by the most diverse services is both an important source of friction among them (Fig. 1–7) and a key to understanding their past and future evolution. This theme is developed further in Section 4 and later sections. First, however, we will show how these same facilities happened to become the backbone of broadcasting, hence a vital factor in bringing both the printed and the electronic news to every citizen.

(Text continues on page 42 . . .)

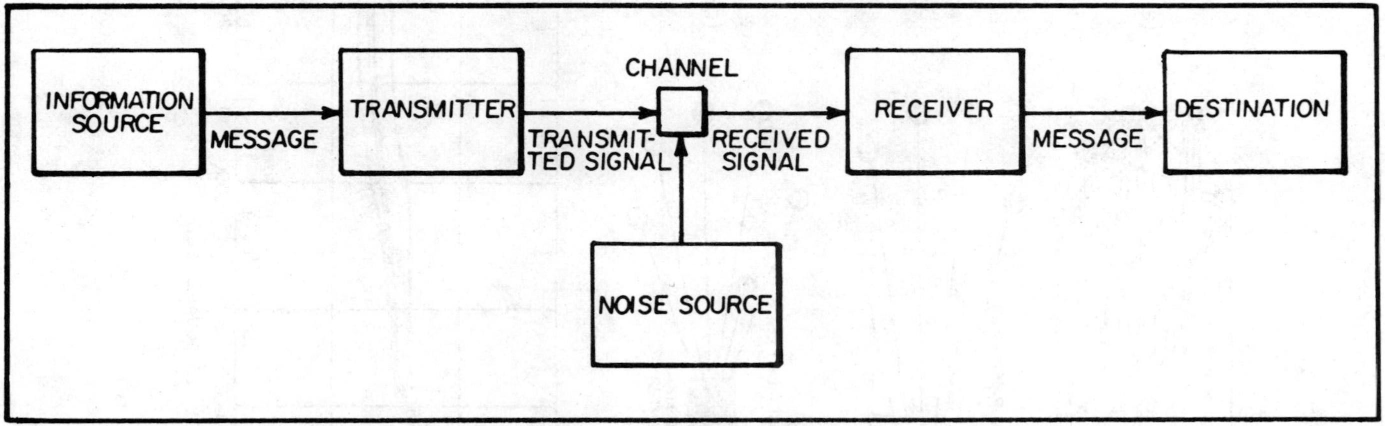

Figure 1–11. A Model Telecommunications Channel

This block diagram was proposed by Claude E. Shannon in his influential 1948 paper "The Mathematical Theory of Communication," which laid the theoretical foundation for digital transmission of information. His work helped define the limits that noise puts on effective information transmission rates as well as practical methods for encoding messages coming from a source into suitable transmission codes. News services use private lines in part because these can be conditioned to minimize noise and therefore give higher quality transmission than random links switched together to make a path from receiver to transmitter.

Source: John R. Pierce, "Communication," *Scientific American*, Vol. 227, No. 3, September 1972, p. 32.

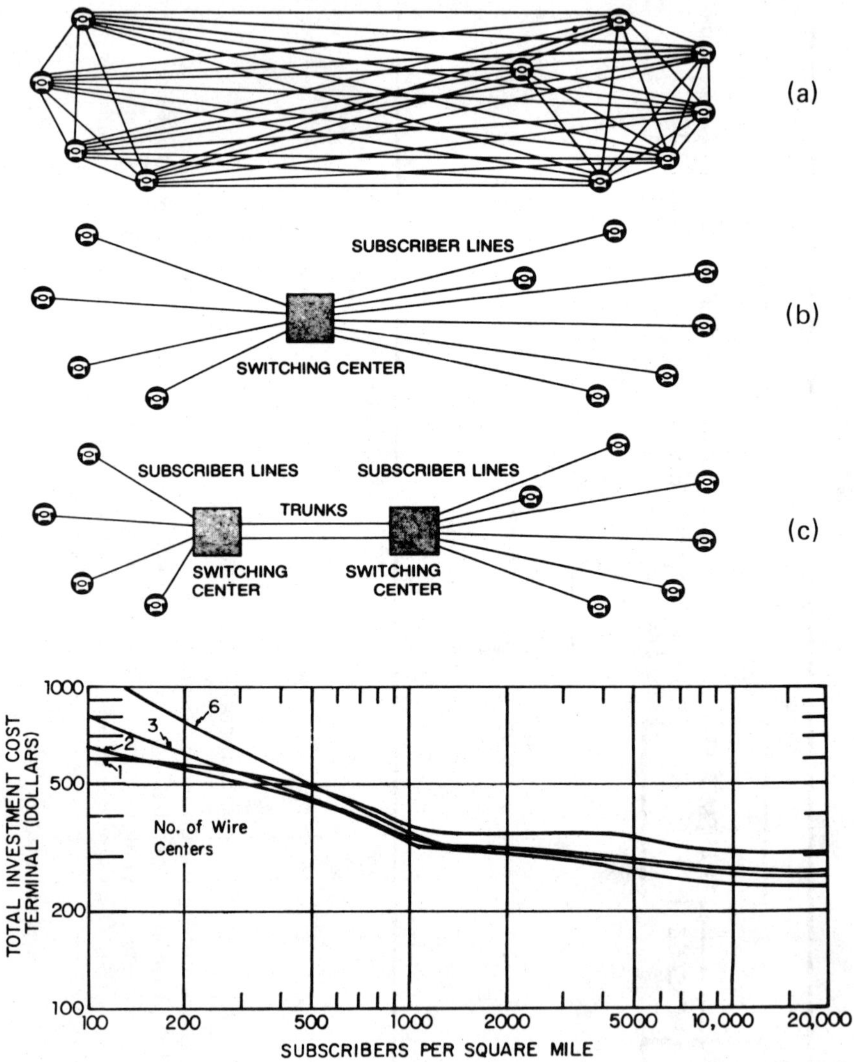
SUBSCRIBER LINES
SWITCHING CENTER
SUBSCRIBER LINES
SUBSCRIBER LINES
TRUNKS
SWITCHING CENTER
SWITCHING CENTER
TOTAL INVESTMENT COST TERMINAL (DOLLARS)
1000
500
200
100
6
3
2
1
No. of Wire Centers
100
200
500
1000
2000
5000
10,000
20,000
SUBSCRIBERS PER SQUARE MILE

(a)

(b)

(c)

(d)

Figure 1–12. Switching and the Practical Interconnection of Telephones

Direct connection between all the telephones of a communication network requires a large number of telephone lines. For example, ten telephones linked in this way would require 45 bidirectional lines (a). A switching center installed between them reduces the number of lines to 10 (b). If the telephones are separated by some distance, trunks connecting the local switching centers reduce the length of each of the individual subscriber lines and trunk capacity is further tailored to the assumption that only a fraction of all telephones are in use at any given time (c). While topologically accurate, (a)–(c) do not account for population densities. Under certain circumstances (d), one switching center is most economical for less than 150 subscribers per square mile, but two are better in the range of 150 to 2,500 subscribers per square mile. Accurate estimates of such factors are important for economic viability.

Sources: (a), (b), (c): Hiroshi Inose, "Communication Networks," *Scientific American*, Vol. 227, No. 3, September 1972, p. 118.

(d) George Mandanis et al., *The Domestic Telecommunications Industry: Economic Behavior, Competition, and Public Policy* (San Rafael, Calif.: Systems Applications, Inc.), May 10, 1974, p. 101.

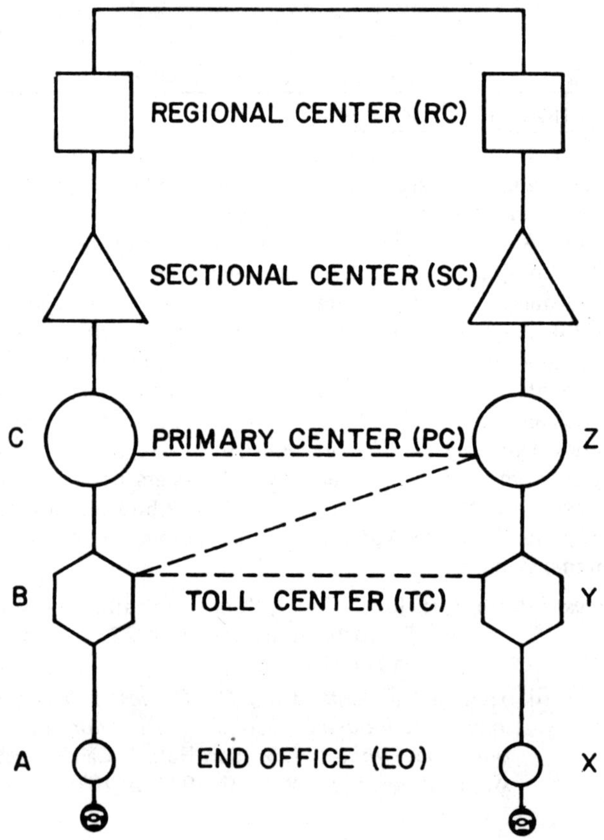
REGIONAL CENTER (RC)
SECTIONAL CENTER (SC)
C
PRIMARY CENTER (PC)
Z
B
TOLL CENTER (TC)
Y
A
END OFFICE (EO)
X

(a)

Figure 1–13. The Hierarchical Network

In actual practice, the principles sketched in Figure 1–12 are elaborated into a network comprising a hierarchy of switches, not all of which are connected to every other, except for the twelve regional centers serving the United States and Canada, each of which has a trunk to every other. Any North American telephone may therefore be connected to any other through at most nine links (a). Where traffic is heavy, additional trunks (dashed lines) are provided and terminals, as well as switching systems, may be directly connected higher up in the hierarchy, omitting some of the intermediate levels shown. Normally, a call from A to X would follow the three-link route A B Y X; however, if B Y should be blocked, the alternate route A B Z Y X will be tried. The ultimate resort is a nine-link route via regional centers. The overall pattern of connections is suggested by (b). The schematic lines in (b) do not necessarily reflect geographical routing, as illustrated in (c), although regions, sections, etc. generally correspond to contiguous geographic areas. Switches of different hierarchical rank may also happen to be located in the same building. In 1975, 700,000 intercity trunks and 2,000 long distance switching systems served the United States and Canada.

Source: (b) Based on figure by Hiroshi Inose, "Communication Networks," *Scientific American*, Vol. 227, No. 3, September 1972, p. 119, and on American Telephone and Telegraph Company, *Notes on Distance Dialing, 1975.*

Figure 1–13. cont.

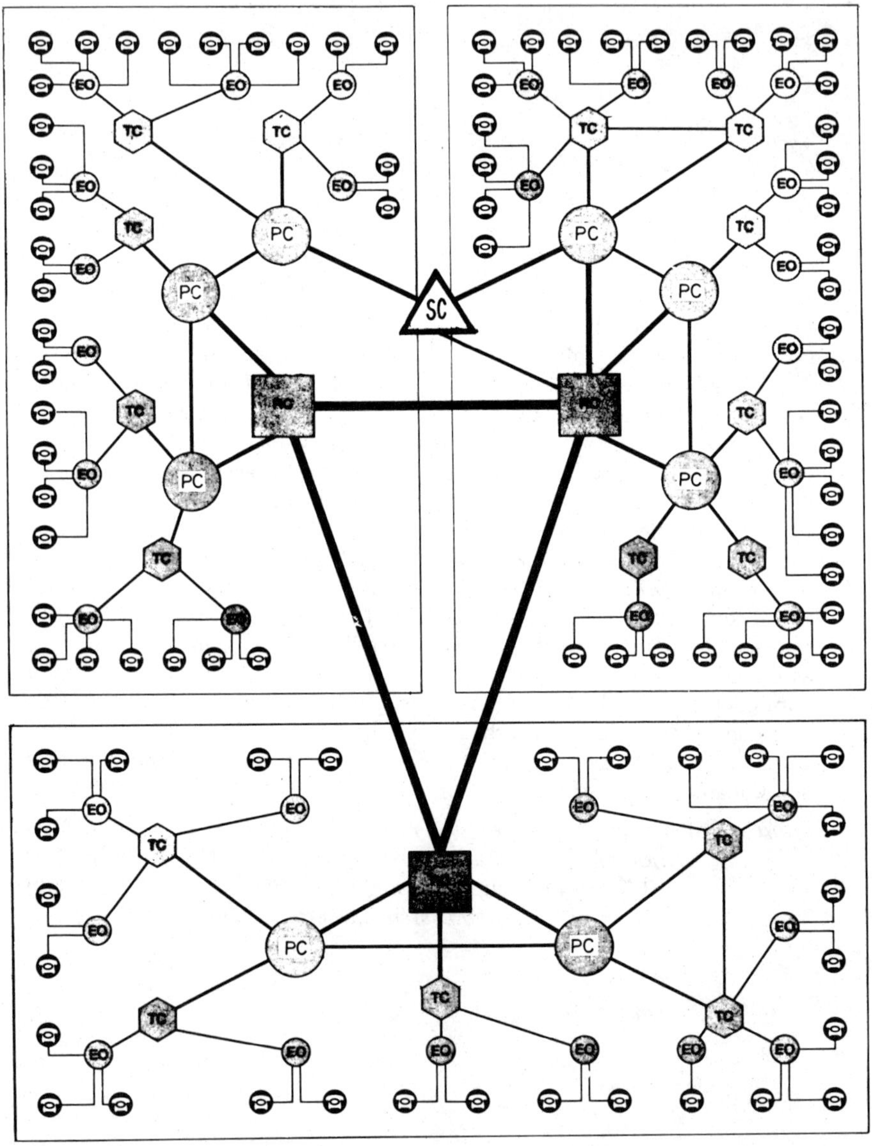

(b)

Figure 1–13. cont.

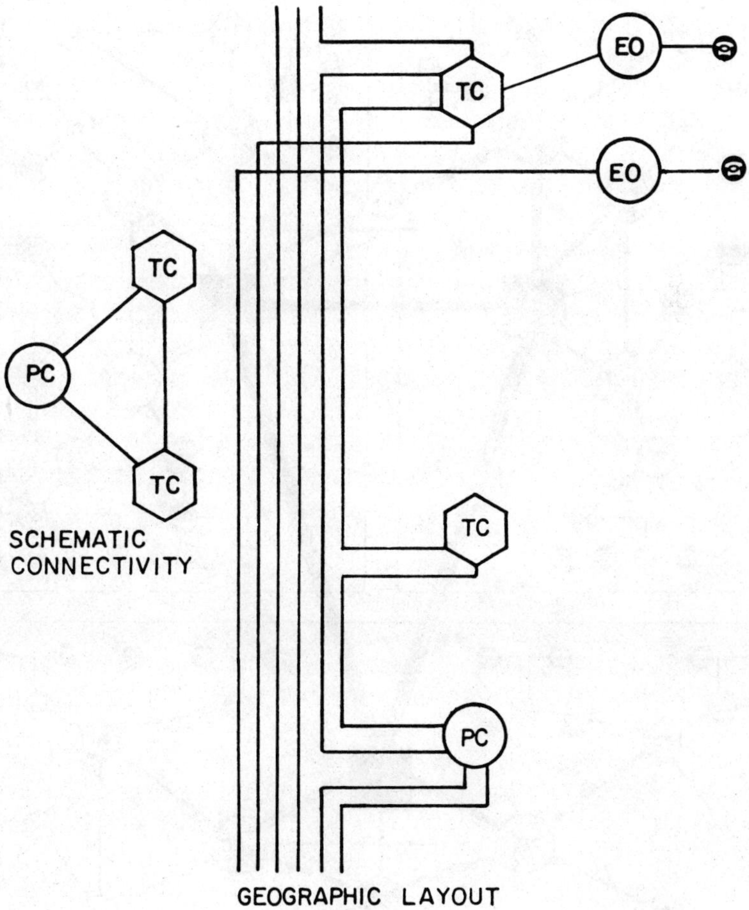

(c)

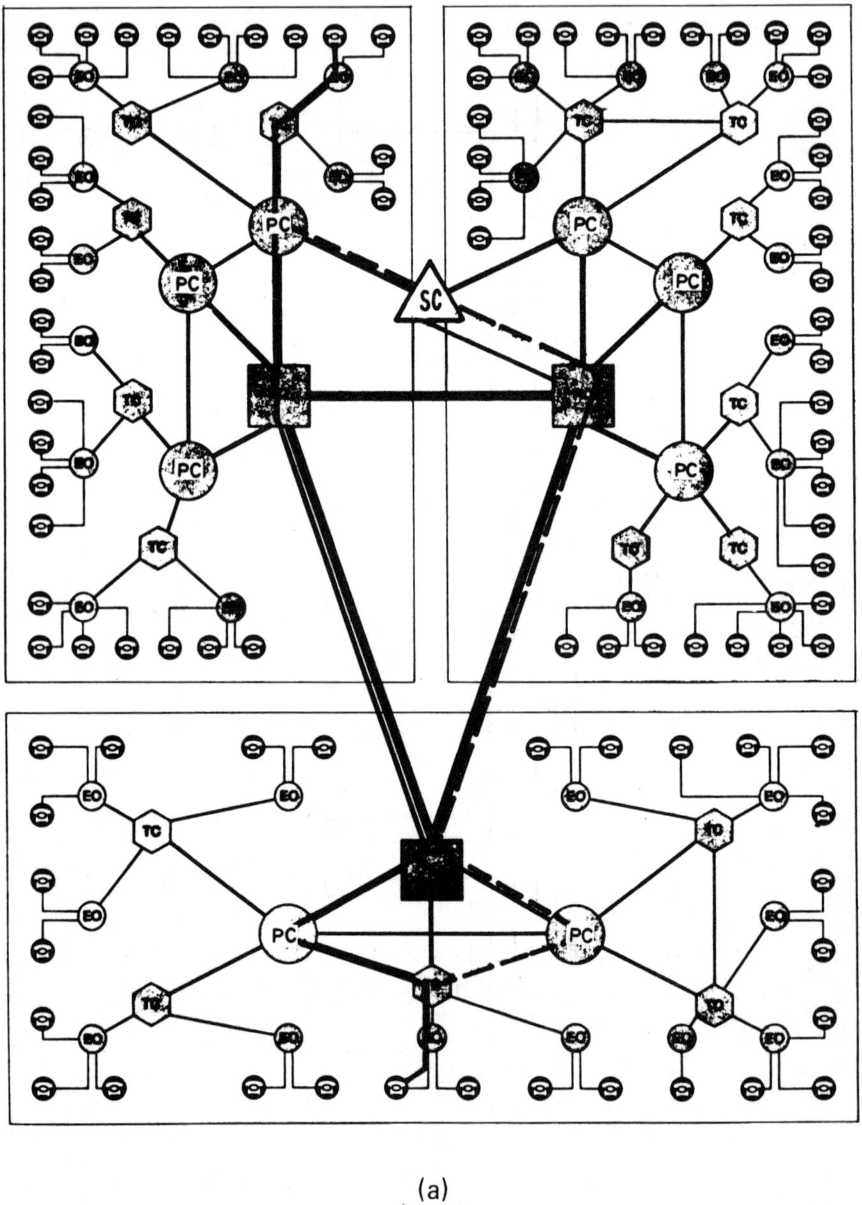

(a)

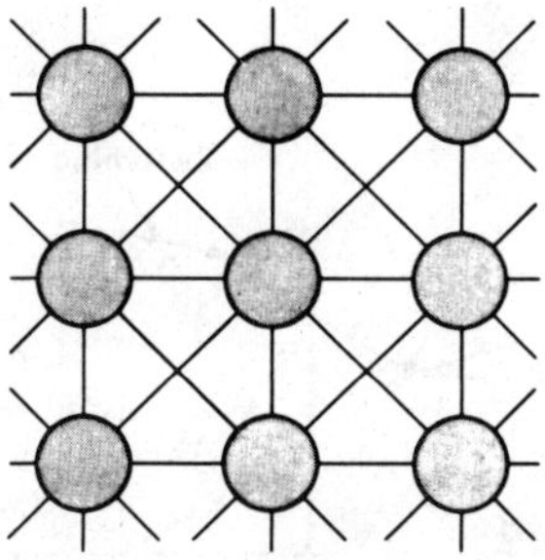

(b)

Figure 1–14. Alternate Routing and Network Reliability

(a) Alternate routing in a hierarchical network is more economical than full connectivity, yet also helps ward off blockage due to ordinary peaks of demand or limited outage due to failure of a cable that might be plowed up by a bulldozer. Extraordinary demand peaks, as during a hurricane, may keep the network so busy hunting for alternate routes that it would block altogether without the manual override function of network management. Major physical damage from accidents or sabotage requires automatic or manual changes in actual physical routes and eventual restoral of damaged facilities.

(b) The military AUTOVON network puts greater emphasis on the "polygrid" configuration which uses more physical links with diversified physical routing and less on the attractive economies of the "star" configurations in the hierarchical network.

Because of alternate routing, paths and transmission quality will vary from call to call. How much alternate routing or how many alternate physical facilities are to be provided by whom and who might bear the costs are among the questions that arise when choosing between monopoly and competition or between regulated and free markets in compunications.

Source: (b): Hiroshi Inose, "Communication Networks," *Scientific American*, Vol. 227, No. 3, p. 127.

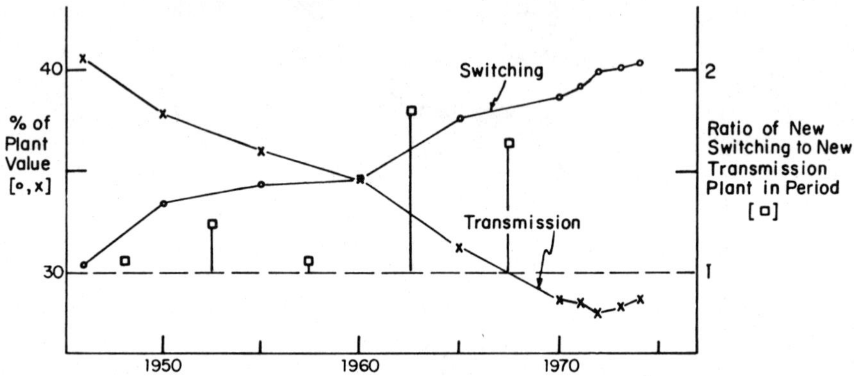

40
30
% of
Plant
Value
[o, x]
Switching
Transmission
2
1
Ratio of New
Switching to New
Transmission
Plant in Period
[□]
1950
1960
1970

(a)

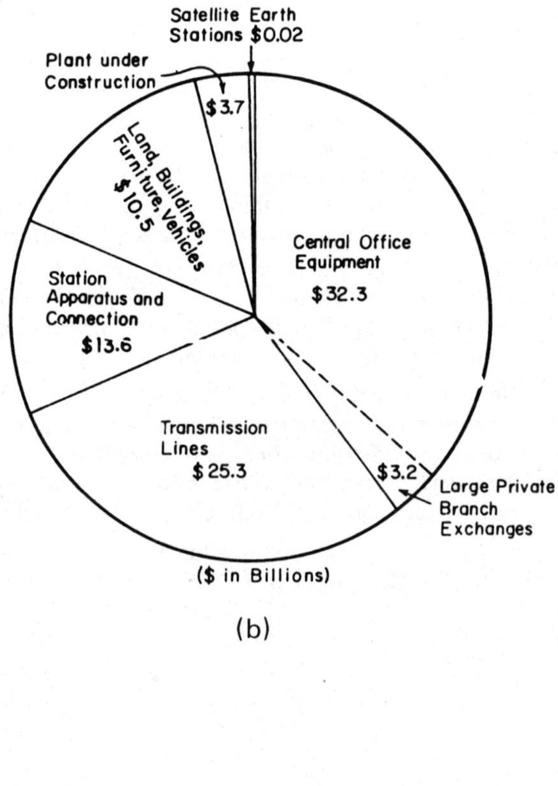

Satellite Earth
Stations $0.02
Plant under
Construction
$3.7
Land, Buildings,
Furniture, Vehicles
$10.5
Central Office
Equipment
$32.3
Station
Apparatus and
Connection
$13.6
Transmission
Lines
$25.3
$3.2
Large Private
Branch
Exchanges
($ in Billions)

(b)

Figure 1–15. Proportions of Switching and Transmission Plant

(a) The reversal in 1960 of the proportions of investment in switching (central office equipment) and in transmission is foreshadowed by an earlier—and continuing—change in the relative rates of investment.

(b) By 1974, switching accounted for $35.5 billion or 40.1 percent, and transmission for $25.3 billion or 28.6 percent of plant value.

Given the dramatic decreases in cost of transmission (see Figure 1–8), it is plain why establishing transmission facilities without switching would appear increasingly attractive, thus accounting in part for the emergence of Specialized Common Carriers (SCCs) in 1970s.

The Uniform System of Accounts, from which the data are drawn, includes such items as multiplex apparatus and covers for transmission power apparatus among central office equipment. Engineers might consider such apparatus as part of the transmission lines. The difference between switching and transmission plant may therefore be somewhat overstated. In practice, these differing perceptions of accountants and engineers complicate the economists' theoretical ideas of costs.

Source: Federal Communications Commission, *Statistics of Common Carriers*, Table 12, 1974 and preceding years.

3. THE BACKBONE OF BROADCASTING

The prevailing linkages among telecommunications, radio, television, and hence the news (Fig. 1–16) are far from being technological imperatives [7]. Actually, the present structure of radio and television networks grew out of settlements of patent conflicts perpetuated by government policy decisions. The U.S. Navy wanted American radio developments to be controlled by domestic organizations in the interest of military and trade security. These concerns were satisfied by 1919 when G.E., the owner of important radio patents, formed the Radio Corporation of America. RCA then took over the assets of the American subsidiary of the British Marconi Wireless Telegraph Company. By 1921, AT&T, United Fruit, and Westinghouse, also patent holders, had joined G.E. in patent cross-licensing arrangements with the fledgling.

Subsequent rapid development of radio broadcasting brought RCA into conflict with AT&T, G.E., and Westinghouse. At issue were not only patent rights and liabilities, but also the structure of the broadcasting industry. In the early 1920s, AT&T aimed to own and operate broadcasting stations as extensions of the telephone network. Broadcasting associations were to be formed by prominent and interested community leaders who would provide all the programs. As with telephone conversations, the associations would pay AT&T for the expense of operating the stations plus a fair return on AT&T's investment. AT&T's plan called for 38 stations to be connected by long distance lines so that a speaker at any one station might address the nation.

When AT&T had trouble buying radio stations and selling time, it tried to abet its aims through its control of both telephone links and transmitter patents. As early as 1922, AT&T had used telephone lines from Stagg Field in Chicago to New York City to describe a Princeton-Chicago football game live over its New York station. Such telephone links were denied to non-Bell broadcasting stations. When this effort failed, Bell embarked on an alternate plan calling for stations equipped with AT&T products to be linked into the long lines network for a license fee. By 1924, the radio network reached from coast to coast. The linked stations were precursors of today's radio and TV network affiliates.

AT&T's dominance over telephone lines led others to seek different ways to link up their stations. Westinghouse, for example, experimented with short wave radio links. Such technological developments, competing with AT&T's control, exacerbated existing conflicts and pressured patent agreements. The result was a new "final" settlement in 1926. AT&T, G.E., RCA, and Westinghouse agreed to a division of roles calling for (among other things) AT&T to sell its broadcasting stations to RCA in return for RCA's promise to use AT&T long lines to link them up. And the National Broadcasting Company, owned 50/30/20 percent by RCA, G.E., and Westinghouse, respectively, was formed to operate the stations. This agreement stuck, and basically still sticks. Later developments were embillishments on this basic theme. When the Columbia Broadcasting System was started as an alternative to NBC, only the programs were different; the interconnections were Bell. Even the antitrust action that split NBC's Red and Blue networks into the present NBC and ABC networks did little to disturb Bell's central position. And when television arrived, it too was plugged into the long lines network with scarcely a thought that it might be otherwise.

The legal framework for radio and television to date is still a portion of the Communications Act of 1934, which embodies the Radio Act of 1927 with but minor changes. As such, the Act of 1934 purports to be based on the premise that broadcasting is a local responsibility exercised on an individual basis by local station licensees [8]. But by 1927, the network structure was in place, and by 1934 it was clear that nearly all the responsibility for producing programming—that is, for keeping the local licensees in business—rested with the networks [9]. Thus, while there may be diversity in the ownership of stations, and hence in the distribution of their revenues, TV program content continues to be dominated by the three major networks.

Diverse technological alternatives continue to present themselves—and to permit alternative policies. In the early 1930s, for example, station WLW in Cincinnati was permitted to operate and "experimental" 500,000 watt "superpower" radio transmitter that blanketed the country—at least at under the favorable conditions of nighttime transmission. But the editorial dictates of the station's owner, Powel Crosley, Jr., were adamantly antilabor. Crosley's one-sided editorial broadcasts led both U.S. Senator Burton K. Wheeler and Federal Communications Commissioner George Henry Payne to see superpower as a potentially dangerous political weapon. By 1938, the Senate had adopted Wheeler's resolution opposing the use of any transmitting power over 50,000 watts, even though fifteen other radio stations were pressing the FCC to approve 500,000 watt transmitters. Claims that such high power was needed as a defense measure failed to catch on. In 1939, the FCC ended WLW's use of 500,000 watts.

Thus by 1940, superpower had been put aside and the telephone network structure of radio broadcasting reaffirmed. However, the FCC's decision to limit American radio power to 50,000 watts did not mean that higher powers were not technically feasible. In Mexico, for example, several superpower stations broadcast in Spanish and English, at nighttime reaching much of the Western Hemisphere. It has been proposed that the regulations of the International Telecommunication Union be revised to allow 2-*mega*watt stations. And, late in 1975, the Federal Communications Commission opened yet another inquiry into the matter of "superpower" for some American radio stations [10]. The Commission itself reported that the objections to the contemplated massive power increases are primarily based on grounds of competitive impact and superpower licensee political power. Moreover, the Commission indicated, the diversion of nighttime radio broadcast audiences to television "may have made the considerable investment in the necessary facilities so unattractive from the economic standpoint that the [superpower] question [may now be] of only academic interest" [11]. In other words, network radio may as a practical matter continue because of network television.

FM radio broadcasting, another alternative to AM radio networking, also suffered setbacks at the hands of the FCC. Developed commercially in the late 1930s, FM broadcasting promised many advantages over existing AM radio, including a static-free signal and the opportunity for increased competition in broadcasting by new stations in a new frequency band; it also presented a challenge to network control of programming. World War II, however, interrupted the FCC's five-year experiment in FM broadcasting, stopping the production of receivers and resulting in a freeze on license grants. By 1944, only 47 FM stations were on the air, serving only about 500,000 FM receivers.

Although great expectations were expressed for FM as "radio's second chance," it soon found itself in direct competition with television for frequencies, advertisers, and financing. The development of television was strongly backed by traditional AM broadcasters and by manufacturers, such as RCA, committed to the speedy development of TV and strongly influential at the FCC and in the Congress. During World War II, the FCC began to reconsider its allocation of the 43–50 megahertz band to FM. Contrary to the recommendation of its own Radio Technical Planning Board, the FCC began to justify moving FM to a higher frequency with highly technical secret testimony about sunspot interferences with FM broadcasting.

On June 27, 1945, pressured by the promised availability of transmitters and receivers soon after the anticipated quick end to the war, the FCC finally decided to uproot FM and move it to its present location in the 100 megahertz band. Although this decision was allegedly technical, the FCC subsequently assigned FM's old frequencies first to TV, which is much more susceptible to interference than is FM, and then to land-mobile services like fire and police radio, where freedom from interference is a crucial necessity. The FCC's decision to move FM and to force the obsolescence of 500,000 receivers dealt FM broadcasting a setback, from which is it beginning to recover only today [12]. In effect, national networks' control of broadcasting had been reinforced.

Thus, through half a century of unprecedented technological change, depression, New Deal, and hot and cold war, broadcasting has retained the structure it was given by RCA and AT&T in 1926. In this structure, the broadcasting networks rely heavily on the telephone company's facilities to get programs from the scene of an event to a distant broadcast antenna, and to link broadcast stations together. These links provide the nationwide program coverage that brings Presidents, astronauts, and "All In the Family" into almost every home—and that also generates the cash and economies of scale to develop the programming that is so attractive to national and local advertisers alike.

The traditional structure of the television networks was never a technological imperative, nor is it so now. The most recent challenger is cable television (Fig. 1–17), hanging on both the telephone and the television networks and with 4.4 percent of cable systems owned by telephone companies and 36.9 percent by broadcast interests as of June 1974 [13]. CATV uses the coaxial cable transmission medium that has long been a staple of long distance telephone transmissions; it uses the spectrum characteristic of broadcasting; and, as of early 1975, it was using terrestrial microwave and looking to satellite systems as alternatives to telephone company networking. The economics and the politics of cable are fascinating, but, as of early 1976, CATV served mainly to redistribute broadcast programs. It remains possible for CATV to emerge as a major challenger to broadcasting and telephony—but that is another story.

Having established that radio and TV, hence electronic journalism, depend on the telecommunications network as much as newspapers do, we turn next to examine the characteristics of this dependence in some detail, particularly in terms of the use of the private line services, where pricing controversies are particularly sharp.

(Text continues on page 50 . . .)

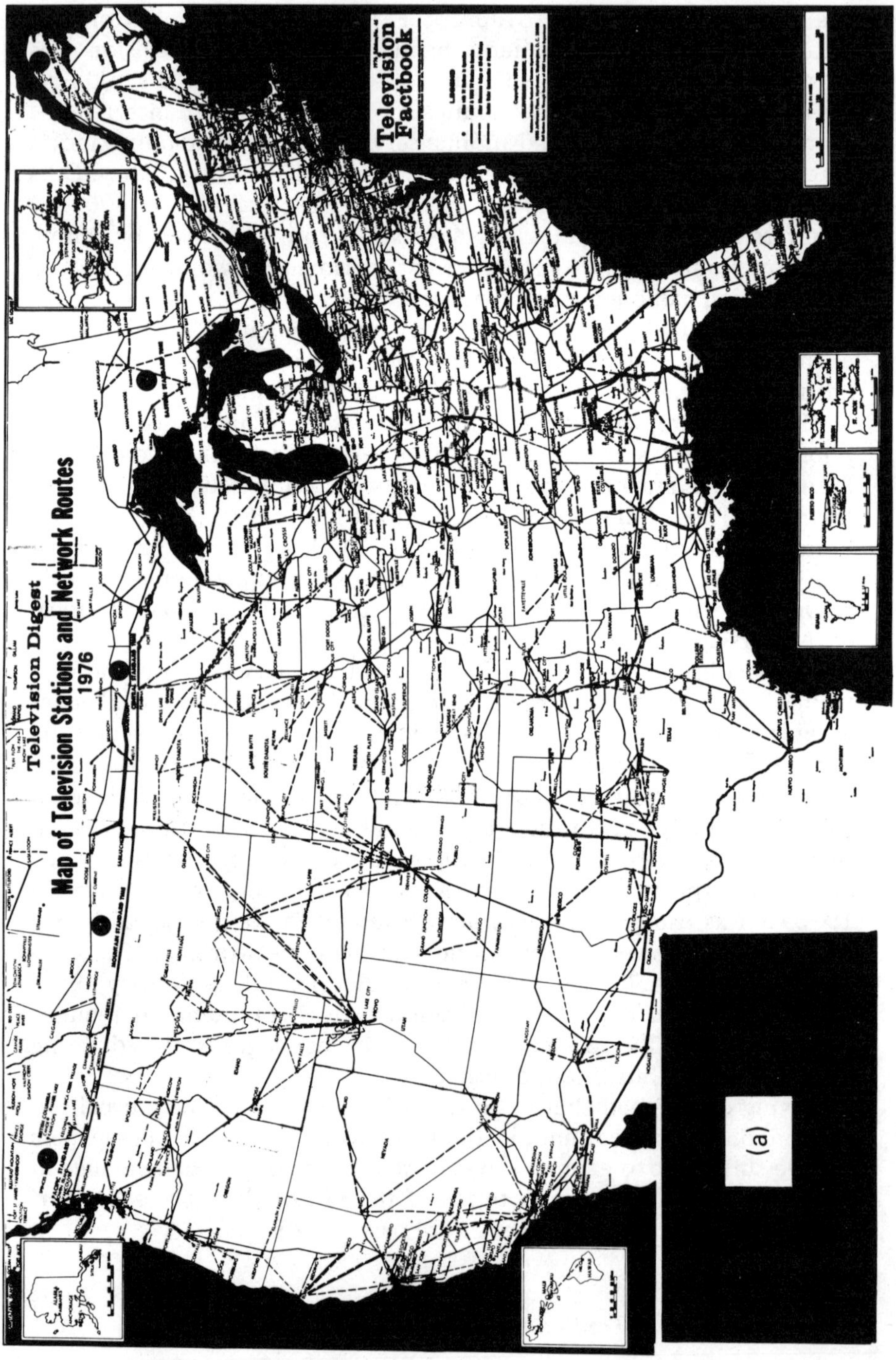
Television Digest
Map of Television Stations and Network Routes
1976
Television Factbook
(a)

(b)

Figure 1–16. The Link-Up Between Telecommunications and Broadcasting

(a) Television networks have long relied on the telecommunications system for the gathering and dissemination of news and other programs. For their ordinary business purposes, TV networks use the switched telephone network like everyone else. But for economical, high quality program transmission, they lease channels (private lines) that are dedicated to their sole use, even though these channels may be derived from equipment shared with other services using the telecommunications network. Such equipment includes physical transmission facilities and other plant, but generally not switching systems.

(b) A typical TV channel has a 6 megahertz bandwidth, the equivalent of up to 2,000 ordinary telephone circuits. In the United States, telephony and broadcasting are governed by distinct legal codes—respectively, Titles II and III of the Communication Act of 1934. That the distinction is growing untenable was recognized by the FCC in a May 1974 decision on mobile radiotelephones where it stated that "In this instance, the radio services we are establishing fall somewhere between what Congress had in mind framing the provisions of Title II and what it intended to cover under Title III of the act" (46 FCC 2d 763 (1974), Docket No. 18262, *Second Report and Order*).

Source: (a) Television Digest, Inc. (Washington, D.C.), *Television Factbook*, 1976 Edition, No. 45. Reprinted with permission of Television Digest, Inc.

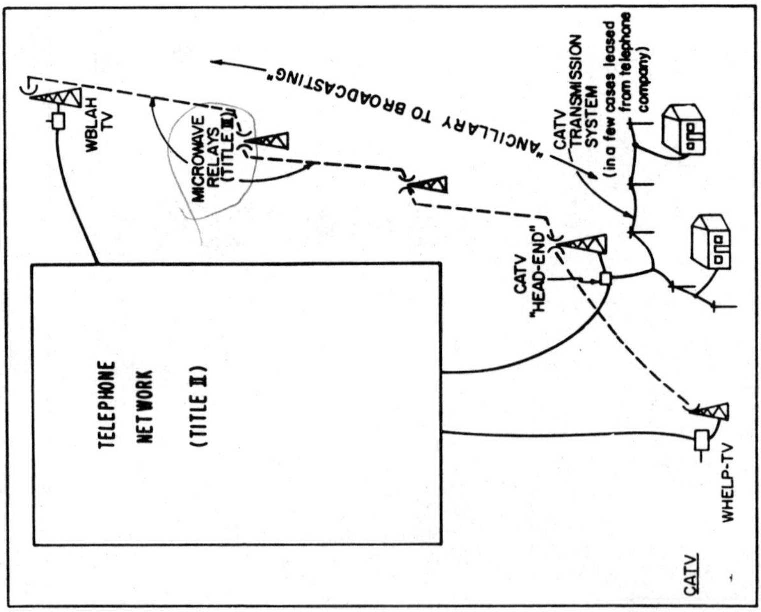

(a)

YEAR	OPERATING SYSTEMS	TOTAL SUBSCRIBERS	SUBSCRIBERS ANNUAL GROWTH RATE (%)
1952	70	14,000	
1953	150	30,000	114
1954	300	65,000	116
1955	400	150,000	130
1956	450	300,000	100
1957	500	350,000	16
1958	525	450,000	28
1959	560	550,000	22
1960	640	650,000	18
1961	700	725,000	11
1962	800	850,000	17
1963	1,000	950,000	12
1964	1,200	1,085,000	14
1965	1,325	1,275,000	18
1966	1,570	1,575,000	24
1967	1,770	2,100,000	33
1968	2,000	2,800,000	33
1969	2,260	3,600,000	29
1970	2,490	4,500,000	25
1971	2,639	5,300,000	18
1972	2,841	6,000,000	13
1973	2,991	7,300,000	22
1974	3,158	8,700,000	19

(b)

Figure 1–17. Cable Television and Telecommunications

(a) Cable television (CATV) systems retransmit local broadcast signals over coaxial cables to subscribers in their franchise areas who might otherwise have difficulty receiving clear signals on their own sets. They also import signals of distant TV broadcasters. The broadcast signals are captured by an antenna at a good location or imported via microwave or other relay facilities. Terrestial microwave relay facilities include CARS, the Community Antenna Relay Service, authorized by the FCC under Title III of the Communications Act as a special service competing with telephone companies, not as a common carriage service under Title II; satellite relay facilities are generally operated as common carriers under diverse ownerships and under regulation conditions that are still in a state of flux.

Retransmission of distant signals has led to bitter warfare between cable operators and local broadcasters fearing competition from the imported signals. CATV's potential, so far unrealized, for local distribution of nonbroadcast signals in competition with telephone companies led these companies to restrict the attachment of CATV cables to telephone poles. A 1967 pole rental agreement stipulated that "under no conditions will the attachment be used to send program material to a single customer or to any small group of individuals" (South Hadley, Mass., and New England Telephone and Telegraph Company, *Agreement*, September 7, 1967, p. 2). This practice was discontinued in 1969 and the FCC has prohibited telephone companies from owning CATV systems in their own operating areas. The nature and extent of FCC jurisdiction over CATV is itself still a controversial matter. As of 1975, overall jurisdiction rested neither on Title II nor Title III, but on two close Supreme Court decisions supporting FCC regulation of CATV insofar as that service is "ancillary to broadcasting." This jurisdiction is further shared with states and municipalities. In February 1976, the U.S. Court of Appeals for the District of Columbia Circuit ruled that the FCC may not preempt state jurisdiction over two-way, point-to-point, nonvideo communications via cable systems (36 R R 2d 393).

(b) The size of the CATV industry is shown as of January 1 of each year.

Source: TV Fact Book, 1973–74 Edition, No. 43 (Washington, D.C.: TV Digest, Inc.), p. 84-a and personal communication with TV Digest Inc.

4. PRIVATE LINE SERVICES AND THE NEWS

Whether by wire, coaxial cable, or microwave radio, television producers, directors, and engineers usually demand high channel quality to gather and distribute TV programs, including the news. Already in 1923, when engineers tried telephone wires to link radio stations WEAF in New York and WNAC in Boston for simultaneous broadcasts, they realized that getting transmissions of high enough quality between the two cities meant dedicating wires—private lines—to that purpose [14].

Generally, this remains true today for both radio and television. While channels so dedicated share equipment with other services, including the general dial-up services, dedicated channels are not dialed up at all. Rather, in order to meet the required quality standards, they must be set up and specially adjusted in advance. Calling the same place very often or very long is often the main reason for using private lines. For example, press services like UPI feed continuous streams of information to newspapers and broadcasters over private lines that cost UPI less than holding a dialed-up connection open. Thus it is economical for UPI to use private lines to interconnect its subscribers, its New York headquarters, major field offices throughout the world, and several switching centers of its own (Fig. 1–18).

In private line services, the communications common carriers supply users with communications paths all their own, each of them conceptually similar to the first telephone installation wherein a pair of wires connected Alexander Graham Bell in one room with one instrument with Mr. Watson in another room with another instrument. Today, many different grades and prices of private line services are provided to transmit conversations, data or broadcast programs. These private lines are dedicated paths through physical transmission facilities common to other services.

News services describe themselves as sensitive to the prices of private line services even though telecommunications costs are not the principal cost element for a newspaper or television enterprise as a whole. Thus, while the networks claim that newsworthiness is the fundamental basis for determining whether to get and report a given story, it is also undeniable that program channels to get a story "live" from remote locations to news headquarters are not free. In some cases, if the price is too high compared with the producer's opinion of the story's newsworthiness, it may be expedient to drop the story altogether or else to tape it on location and then use cheaper physical transportation with attendant loss of timeliness. And private lines are the facilities that distribute all the news—and all other current programming—to network affiliates.

Cost, of course, is not the only mundane factor that affects the availability of channels and hence the content of the news. During the Mideast War of October 1973, about three times more television news came from Tel Aviv and Jerusalem than from Cairo and Damascus. Sig Mickelson, a former president of CBS News, noted that "superficial analysis might suggest that this phenomenon was caused by an emotional commitment to the Israeli cause." He points, however, to "mechanical, logistical, and governmental restraints that influence the end product as seen on the home screens." Specifically, Israel could feed program material to the Intelsat satellite system through an earth station near Tel Aviv. Neither Egypt nor Syria had earth stations and the station in Jordan was virtually inaccessible. The advantage to news from Israel was a "lead time of 12 to 24 hours . . . sufficient to make the difference between major and secondary treatment, or even no story at all." There were also the additional advantages "of more open government policies, working inside the perimeter of the battlefronts rather than outside (a significant advantage considering the weight and volume of equipment), [and] a common language with many Israeli troops" [15].

Thus, prosaic factors like the prices of private line service and international agreements on satellite links have at least some impact on the origin, timeliness, treatment, and distribution of the television news. The effect of private line prices on the clients of press services like AP and UPI may be more direct, since the latter have both the incentive and the possibility of passing on price increases to their customers. If prices rise high enough, smaller news retailers may be forced to stop using at least some of the several different press services offered. And since the recipients of news from the wire services are also collectively important suppliers of news to them, the effects of such shifts in newsgathering methods on the accuracy, completeness, and diversity of the news could be dramatic. Accordingly, attorneys for news organizations have argued in many regulatory proceedings that the price structure of private line services is a key element in determining what news people get from where. More generally, how and what prices are set for *any* customer is critical in determining "who gets what, when, how"—Harold Laswell's phrase defining politics [16].

The next section therefore goes into some detail about the theory and the practice of determining telecommunications costs and setting telecommunications prices as it affects everyone, and not only the example of the news. The political accommodations described in the next section grew out of many skirmishes over the century since the telephone was invented. How these accommodations threaten to be disturbed by recent changes is the theme of Section 6. The significance of the changed situation for the news is developed in Section 7.

(Text continues on page 54 . . .)

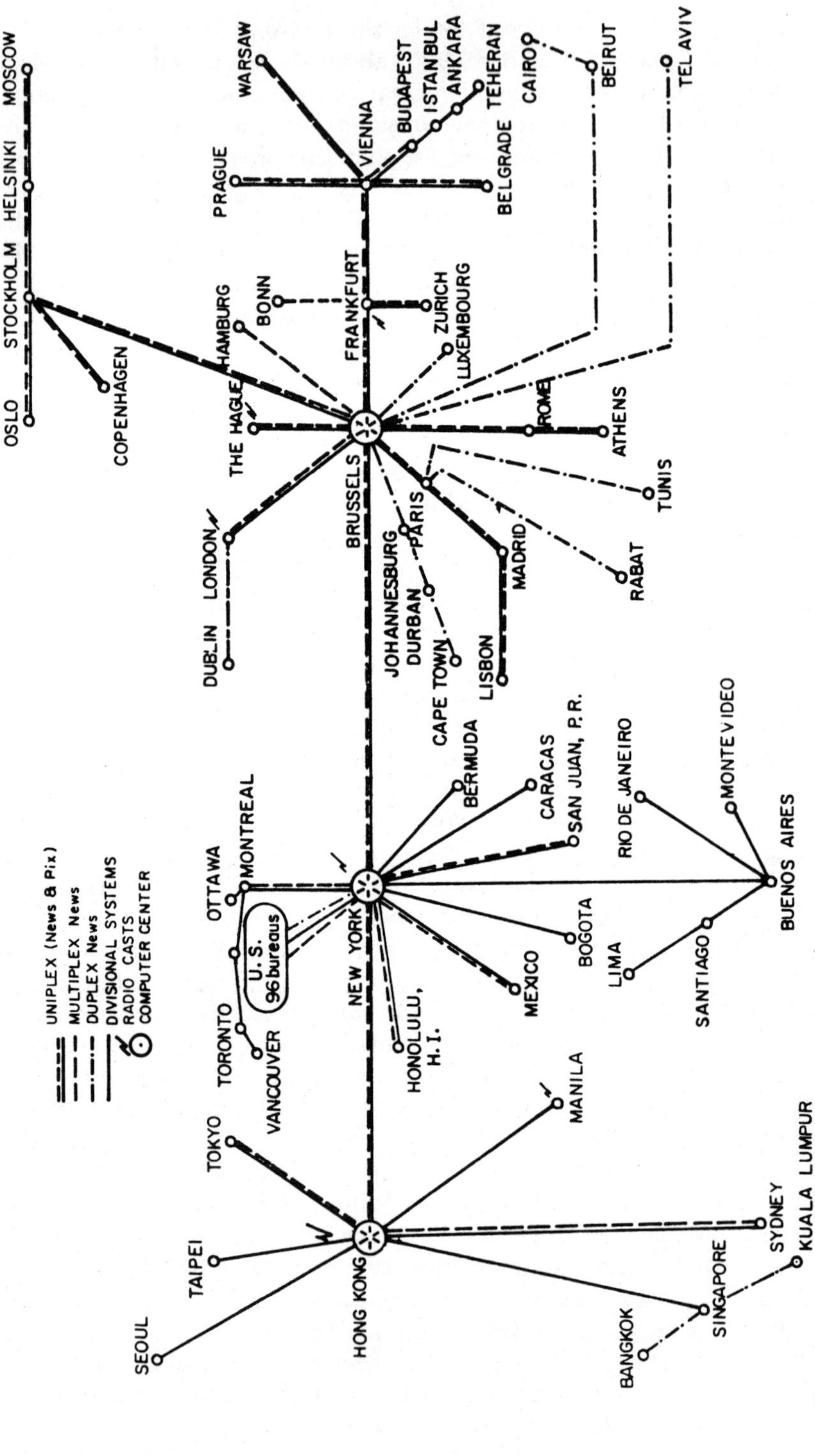
UNIPLEX (News & Pix)
MULTIPLEX News
DUPLEX News
DIVISIONAL SYSTEMS
RADIO CASTS
COMPUTER CENTER
OSLO
STOCKHOLM
HELSINKI
MOSCOW
COPENHAGEN
DUBLIN
LONDON
THE HAGUE
HAMBURG
BONN
PRAGUE
WARSAW
BRUSSELS
FRANKFURT
VIENNA
BUDAPEST
ISTANBUL
ANKARA
TEHERAN
BELGRADE
ZURICH
LUXEMBOURG
ROME
ATHENS
CAIRO
BEIRUT
TEL AVIV
PARIS
MADRID
LISBON
JOHANNESBURG
DURBAN
CAPE TOWN
TUNIS
RABAT
OTTAWA
MONTREAL
TORONTO
VANCOUVER
U.S. 96 bureaus
NEW YORK
HONOLULU, H.I.
BERMUDA
CARACAS
SAN JUAN, P.R.
MEXICO
BOGOTA
LIMA
SANTIAGO
BUENOS AIRES
RIO DE JANEIRO
MONTEVIDEO
TOKYO
MANILA
HONG KONG
TAIPEI
SEOUL
BANGKOK
SINGAPORE
SYDNEY
KUALA LUMPUR

Figure 1–18. UPI's International Private Line Network

UPI's international private line network includes major switching centers in Brussels, Hong Kong, and New York. A national private line network also fans out of New York, where editors use video display terminals connected to computers to write, edit and direct the movement of news copy to subscribers. The diffusion of electronics technology since World War II has given heavy telecommunications users like UPI additional means for reducing their costs. By installing at their own switching centers so-called multiplexing equipment readily available from many sources, they may subdivide a private voice line into 25 to 30 teletypewriter channels. Pricing is such that these channels cost less than they would if leased directly from a common carrier. The brokers, service suppliers, and associations described in Figure 1–7 saw an opportunity to share similar savings with groups of customers unable to afford this opportunity on their own.

Source: United Press International.

5. COSTS, PRICES, AND THE POLITICS OF TELECOMMUNICATIONS

Economic Theory and Practical Costing

Modern economists think that the greatest profit, economic efficiency, and social welfare—as they define these ideas—all coincide when the price of a good or service to consumers and the marginal revenue and marginal cost to producers also coincide, as they do at equilibrium under the theoretical conditions of perfect market competition (Fig. 1–19a). This restates more explicitly Adam Smith's belief in market competition as a mechanism whereby the entrepreneur bent on "only his own gain" is "led by an invisible hand" to promote society's interests.

Even if it ever did or could, telephone service has not fit that mold since 1913. By law, it has become essentially a geographic monopoly with but a single telephone company serving any one area. Consequently, it has been both regulated "in the public interest" and operated "in the public interest"—a motive that Adam Smith scorns when it guides a visible hand, claiming he has "never known much good to be done by those who affected to trade for the public good" [17]. How visible what hands should be remains a major policy question.

According to elementary economic theory, regulated monopoly prices are set in terms illustrated in Figure 1–19b, which assumes that the firm makes a single product at definite costs to maximize profits in meeting a market demand related to the product price in a definite way. In actuality, telephone facilities are used to offer a variety of services broadly classified as shown in Figure 1–20. Some of these services may be substituted for one another to greater or lesser extents, exhibiting, in economist's terms, cross-elasticities of demand. An example is the substitution of private line for dial-up service (MTS) when price and volume warrant it. Others are complementary in that each is needed to use the other. Still others go together optionally.

Private line services encompass not only transmission services, but also such optional services as common control switching arrangements (CCSA) and foreign exchange service (FX). CCSA provides for switching within a customer's private line network, thereby putting at the customer's exclusive disposal what amounts to a miniature version of the whole telephone network. In FX, a private line reaches to a distant exchange, enabling the user to call and be called from telephones at the remote exchange just as if they were local. Economic theory can and does cope conceptually with these and many other complications of the single product model; only the mathematics gets a bit messy. Some complications that arise in practice still lie beyond the scope of economic theory, but there is active research on extensions and modifications of the available theory.

The crucial problems come up in linking theory to practice. Fundamental concepts of physical science, such as mass, velocity, or temperature, nowadays can be linked to reality through repeatable measurements of exquisite accuracy and precision. Not so for cost, not so for demand; not so, that is, for the fundamental concepts of economic theory. And practical difficulties in measuring or estimating are only part of the problem. In Figure 1–15, for example, we noted how arbitrarily the Uniform System of Accounts prescribed by government regulators distinguishes transmission from switching costs. The concept of true economic cost is thus far more tenuously linked to practical measurement than, say, the scientific concept of velocity is to measuring the speed of a car. And even if it were so, we will show that explaining how multiple telecommunications services are priced and how these prices relate to costs unavoidably takes some appeal to power and to politics as well as to the technicalities of technology and economics. Technology sets bounds on our latitude, and economic theory can clarify problems and guide reasoning about choices; but interpretations vary and choice remains discretionary.

As Figures 1–13 and 1–14 show, a dial-up (MTS) call between two points has no unique and foreordained correspondence with the physical facilities used to complete it. A private line, on the other hand, is conceptually unique and distinct. However, it is far from being physically separate from the facilities used for all other services. A "private" pair of wires to an end office hangs on the same poles or runs through the same underground ducts as do "public" wire pairs. At the central office, the equipment that defines private channels in a microwave trunk shares racks, power supplies, space, heat, and light with the equipment defining all other channels on the trunk. Transmission facilities such as radio transmitters, towers, antennas, and the land they stand on are shared by all the many services carried over the same route. FX and CCSA services use, or at least access, switches that also serve MTS.

Since not only physical facilities but also operating and management personnel are shared by many services, there are among these services many common and joint costs. Defining *equitable* relationships between such costs and the prices of distinct services sharing common facilities is among the thorniest and most vexing issues which accounting and economics can describe and analyze but which politics must resolve. In part, this is "merely" a matter of practicality. It is conceivable (and sometimes practical) to trace and to meter the specific facilities used by a call; only the cost of attributing costs by capturing, aggregating, and using the metered information gives pause. But what principle should guide, for example, the treatment of the fixed costs of facilities that are standing by, waiting to be used . . . by what service? Since fixed plant investment is so large ($88.6 billion book value of total communication plant as of 1974) and the variable costs of service are so difficult to track, associating revenues (prices) with costs (Figs. 1–21 and 1–22) is a fundamentally *arbitrary* process.

As is widely the case, management does not know its own costs for sure. The minutes of AT&T's Executive Policy Committee Meeting of November 9, 1972, brought out by Senate hearings in July 1974, report that functional accounting "is urgently needed to provide the companies readier information than current ad hoc studies provide with respect to the cost and profitability of their products and services" [18]. A submission by the United Telephone System to the Federal Communications Commission in 1975 states that "the only way . . . to determine the costs of specific services . . . is through an arbitrary allocation of costs of jointly used facilities . . . United will conduct certain studies . . . to develop specific data concerning costs as a function of exchange size, customer calling patterns, and population density . . . scheduled for completion during . . . 1977" [19].

Costs, Revenue Requirements, and Pricing Policies

The clear and distinct cost curves in the graphs or equations of economic theory are thus matched up to reality only very fuzzily at best. Moreover, whatever the advice that economic theory might offer, in reality the relationship of prices to costs is discretionary within ill defined but broad limits that depend on policy goals.

The obscure but not invisible hands that set telephone prices (Fig. 1–23) proceed roughly as follows. First, regulators determine the telephone companies' revenue requirements. Then individual services are priced by the companies to meet these revenue requirements. In gross terms, revenue requirements are set equal to the "cost" of providing all the regulated services that a telephone company offers. "Cost" in this context is what lawyers call a "term of art," defined roughly as follows [20]:

Annual cost of services = operating expense + depreciation expense + taxes
+ [(plant investment − accrued depreciation)
× (a rate of return on plant)]
= plant investment × annual carrying charge
= *annual revenue requirements*

Each of the items on the right hand side of the equation and each of the components of each of the items, such as "operating expenses," might be the subject of regulatory proceedings—replete with engineers, economists, lawyers, investors, bankers, judges, and politicians—to determine its "proper" level in the regulatory equation.

Of special interest is the rate of return on investment, which is set at percentages (8 to 11 percent overall for AT&T in the early 1970s) somewhat lower than rates in riskier industries, but aimed by law at insuring the financial integrity of the company while allowing for the attraction of new capital. Actual revenues are usually more or less than calculated revenue requirements depending on actual market demand and actual costs.

For a time, telephone companies may tolerate lesser profits or else cut costs rather than ask regulators to approve price increases; regulators tied up in numerous other proceedings before them may tolerate higher revenues for a period; and, in any case, the regulatory process itself is slow moving and subject to protracted challenges in the courts. The effects of these "regulatory lags" on the behavior of regulated firms are hotly debated. Within the bounds of the overall "revenue = cost" equation, there is considerable latitude in setting prices for specific services aimed at particular categories of user demands. Indeed, as illustrated below, there is generally no necessary relationship between the cost of a service and its price.

Telephone company managements usually take the initiative in price changes. Their prices for the various types of service are set forth in published rate schedules or "tariffs" according to management judgments of how the price of a service, times estimated demand, summed over all services, might best approximate the established total revenue requirements, while also meeting regulatory standards. The standard set by the federal statute, for example, is that prices to consumers and returns to producers "shall be just and reasonable," allowing for such classifications of services as the Federal Communications Commission "may decide to be just and reasonable" and for different prices in different classes [21].

Some economists see the "just and reasonable" standard as met when prices are set at the level that would be dictated by perfect market competition, for theory tells them that social resources will be allocated optimally at those prices. In reality, however, there are very few, if any, perfect markets. In telecommunications particularly, prices are set through regulatory proceedings involving adversaries. In that setting, the normative precepts of diverse schools of economic thought are controversial and not generally accepted even when the issues fall within the bounds of current economic theory.

The attributable costs of various services are fuzzy, and shares of common costs are as arbitrary here as everywhere. The textbook demand curve cannot be determined empirically: data about past consumer responses to price changes yield at best a small segment of a demand curve and making marketing projections is at best an art. The practical limitations on applying the explanatory power of economic theory are thus apparent. Even in principle, applying economic theory to a specific situation requires that policy "constraints" be made explicit. To understand actual prices, one must therefore refer to the interests and the powers of the players listed in Figure 1–23 as well as to economics and technology.

For example, the widely used "value-of-service" pricing method (Fig. 1–24) shows how establishing fairness and reasonability is in essence political. In value-of-service pricing, users willing to pay more for a given service are charged more—in this case city dwellers as compared to farmers, businesses as compared to householders—even though it may actually cost more to service those least willing to pay. Value-of-service pricing, as in Figure 1–24a, has been euphemistically (but not without basis in theory) labelled as the intelligent exploitation of differing demand curves, leaving unexplained how these generally unmeasurable curves are known [22]. Of course, the cost curves, which are supposed to intersect the demand curves intelligently, are themselves fuzzy, arbitrary and often unknown.

Implicit in Figure 1–24a is another pricing concept called "rate averaging." As shown in Figure 1–12d, costs are higher where subscribers are sparse, yet prices in Figure 1–24a are lower where fewer subscribers may be reached. Moreover, there is one price for every subscriber in a given area, no matter how short or long his lines are to the nearest central office. In part, this reflects the politically determined policy not to influence residential or business location decisions through pricing of telecommunications services. In part, it separates engineering decisions about the best location of telephone exchanges from price considerations that might otherwise lead to locating an office near some large customer to reduce his price while raising those of others.

Not that basing prices on demand rather than costs is necessarily bad economics that fails to make best use of society's resources. This is most clear cut in Figure 1–24b and c, which illustrate the economic rationale for the price differentials between daytime toll rates and the rates for evening and weekend toll calls. But distinctions among services using the same plant do not always reflect differences as sharp as those among loads at different times of day. What are *like* services (where price differentials are discriminatory) and what are *distinct* services (where price differentials can be just and reasonable) is as hotly argued in practice as what are costs.

The local price structure illustrated in Figure 1–24a prevails in most states. Through its participation in the "separations" process, described next, the Federal Communications Commission accepts it as consistent with its broad statutory mandate "to make available, so far as possible, to all the people of the United States, a rapid, efficient, Nation-wide, and world-wide wire and radio communication service with adequate facilities at reasonable charges" [23]. The telephone companies themselves espouse a goal of universal telephone service. The White House Office of Telecommunications Policy, on the other hand, has advocated cost related pricing for common carrier services "because of [that method's] role in insuring the efficient use of that resource" [24]. Stating that carriers and regulators accept "that certain forms of cross-subsidization are in the public interest," it suggests that "when business service is over priced, the cost of this service may be passed on to the business consumer as a hidden sales tax with typically regressive effects." It concludes that "this is not to say that economic support for certain classes of telephone users is necessarily bad policy, but rather that this form of hidden subsidy leads to economic inefficiencies and poses obstacles to the effective introduction of competition" [25]. We return to such points in Section 8.

Ownership and Jurisdictions: The Influence of Settlements and Separations on Costs and Prices

All this arbitrariness is still but a part of the cost and price story. A tangled overlay of ownerships and of federal, state, and local jurisdictions (Fig. 1–25) adds yet another element to the calculus of the just and the reasonable, and further complicates the costing of various services and the relationship of prices to costs. For example, a telephone call originating in the territory of the independent Souris River Telephone Mutual Aid Corporation in North Dakota may be carried over the facilities of Northwestern Bell Telephone Company, AT&T Long Lines, the New York Telephone Company, and the independent Rochester Telephone Company to reach its destination in Rochester, New York. Since this call crosses state lines, its price comes under the jurisdiction of the Federal Communications Commission. Had the call been within one state, a state agency would have had jurisdiction over its price. The revenues both from intrastate toll calls and from interstate toll calls such as this one are shared among AT&T Long Lines and affected operating companies by an intercompany settlement process.

The overall equation "cost of service = revenue requirement" therefore has many overlapping parts. Overlaying the settlement shares is a jurisdictional separation of costs into interstate and intrastate components that contributes to determining the revenue pools that are shared among companies through settlements (Fig. 1–26). Separation introduces further arbitrariness into the definition of the costs, hence of the prices that are set to meet revenue requirements for each jurisdiction. Since the overall equation holds, and double counting is to be avoided, any increase in the interstate share implies a decrease in the intrastate share and vice-versa. Within each state jurisdiction, there is the further question of allocating costs between intrastate toll and local services. Until 1971, only California regulators separated costs between intrastate toll and local services. New York then adopted the same approach. In all other states, intrastate prices—toll and local—are directly related to costs only through the overall statewide equation. The result is a multiplicity of revenue requirements, each for a bundle of services, principally the bundle of all the services offered within a jurisdiction.

The price for each service in a bundle is then set, as described earlier, within the constraint that price times demand for each service, summed over the services, approximate the revenue requirements for the bundle. As we shall see, the interplay among pricing policies, separations and settlements is an exquisitely arcane political process. At stake is how much of the aggregate cost of local and toll facilities is paid for by users of local services; how much by users of toll services; how much by users of intrastate toll services; how much by users of interstate toll services; as well as the equities of further details of pricing patterns within each of those broad service categories—or, from the viewpoint of the telephone companies, how much in revenues which service can earn.

Separation and settlement distinctions therefore matter to different types of customers, both among themselves and vis-à-vis the telephone companies; they matter to independent telephone companies vis-à-vis the Bell System, to Bell operating companies' management vis-à-vis AT&T Long Lines and corporate management; and they matter to all companies in terms of which legislators or regulators—state or federal—might be best to deal with on revenue requirements, on what services customers might be most willing to pay how much for, and on what tax or other consequences might ensue from how the pie is cut.

The details of how intercompany settlements and jurisdictional separations evolved are manifold and interpretations of their significance differ widely. The trend, however, has been unmistakably toward increasing the proportion of local costs counted toward interstate toll revenue requirements, meaning that users of interstate services pay a larger and larger part of the total bill.

Up to about 1930, jurisdictional cost separations were not a major factor; contractual intercompany settlements were reckoned on other bases, such as current expenses. Costs were reckoned along lines following a fairly natural institutional and technological boundary between then separate local and toll switchboards. Facilities from telephone stations up to and including the local board belonged to local service and those between local boards to toll service. In contemporary terms, this is described as a "board-to-board" principle. Under separations based on such a principle, toll services would bear only the costs of adding toll services on to the "basic" local service, as schematized in Figure 1–21(b2).

In 1930, in *Smith* v. *Illinois Bell Telephone* [26], the United States Supreme Court held that some of the local facilities' costs should be allocated to intrastate and interstate toll services, taking into account relative use of the facilities, since toll services necessarily use some, but hardly all, of the local exchange facilities. This decision signaled the rise—opposed by the Bell system until 1943—of separations based on "station-to-station" concepts, i.e., on including a portion of local facilities costs among those used in determining interstate toll revenue requirements. Results are as schematized in Figure 1–21(b1) or (b3).

Figure 1–27 illustrates the effect of increasing allocations to interstate revenue requirements on the income of an independent telephone company. In 1973, United Telecommunications, the nation's third largest system (Fig. 1–25c), allocated about 50 percent of its local plant investment to toll services [27]. Interstate and intrastate costs, in turn, each account for about one-half the toll share. The growth of the interstate share as separations went from "board-to-board" to "station-to-station" principles is evident from Figure 1–27. It is noteworthy that each dollar of local plant investment added to interstate toll services generates between 25 and 30 cents a year of added interstate revenue requirements. A share of actual interstate revenues proportional to the independent company's share of interstate revenue requirements accrues to it in consideration for the interstate use of local plant. And the actual revenues shared are generated not only by calls involving the 17 percent of all telephones belonging to independents, but also by calls among the 83 percent of telephones served by the Bell System.

The plan governing jurisdictional separations as of late 1975 was developed at meetings held in August 1970 at Lake of the Ozarks, Missouri, by a joint board made up of four representatives of the association of state utilities regulators (NARUC) and three Federal Communications Commissioners. The FCC ratified the plan in October 1970 [28]. The plan shifted another $130 million of costs or revenue requirements, or about 8 percent of AT&T's net income for 1970, from state to interstate services.

The mechanics invented to implement the Ozark Plan include apportioning some kinds of telephone plant according to a "subscriber plant factor." This factor takes into account, among other things, what is described as the deterrent effect on toll calling caused by the fact that toll prices are based on usage, while most basic exchange prices are based on usage-insensitive "flat" rates. The idea is to increase the weight of interstate toll usage relative to that of local and intrastate usage in order to make cost shares come out at a level agreeable to the parties, all the while keeping up the appearance of allocations based on the court-sanctioned concept of relative use of facilities. Although couched in internally consistent logic, this ingenious excursion into a formalistic Wonderland is nonetheless an ex post facto account of a basically political resolution of one facet of the problem of arbitrarily allocating common costs.

Under the Ozark Plan, the proportion of local facilities costs ascribed to interstate toll usage tends to increase as the usage of interstate MTS and WATS increases, since the subscriber plant factor is sensitive to the number of calls that are made in these services. However, there is no metering of traffic carried in the private line services and no influence of private line traffic volume on the subscriber plant factor. This technicality reduces the share of plant costs allocated to interstate services whenever a customer shifts from MTS to WATS to private line services or to dial-up services like the contested Execunet service proposed by Microwave Communications, Inc., one of the organizations (discussed in Section 6) that the FCC calls "specialized" and AT&T "other" common carriers.

In the net effect, the Ozark Plan allocates 18 to 25 percent of the costs of local facilities to the interstate revenue requirements. Each dollar of costs so allocated yields at least 25 to 30 cents a year of interstate revenues to a local operating company. Thus, at present, the fortunes of independent companies depend significantly on those of the nationwide interstate traffic generated by actors other than themselves. Whether fought out over cost allocations or on some other basis, this sharing of interstate revenues has been at issue between the independents and the Bell System ever since long distance telephone began; to a lesser degree, it affects relationships between the managements of Bell System operating companies and the parent company.

What does the separation process mean to users and to regulators? Historically, the "short haul . . . state toll business has inherently a high cost structure per message relative to interstate traffic" (Fig. 1–28) [29]. Interstate traffic, because of major advances in long haul transmission technologies (see Fig. 1–8), "is relatively low cost per message with relatively high revenue per message" [30]. Moreover, the history of toll demand lends credence to a longstanding assumption that it is more elastic than local demand (Figs. 1–29 and 1–8d). Therefore, keeping interstate costs (revenue requirements) relatively low would tend to increase total revenues to AT&T from the long haul "cream" of its services and also further AT&T's aim to solidify its monopoly of that cream. The benefits to users of interstate toll services are clear. On the other hand, prices for intrastate toll calls have generally been higher than those for interstate calls (the so-called toll disparity), reflecting in part the cost differentials shown in Figure 1–28. In addition, however, meeting intrastate revenue requirements means setting pricing policy as to how much revenue is to be generated by local services and how much by intrastate toll services.

Historically, the tug-of-war among state regulators, local telephone companies, business or residential users, and urban or rural users has led to pricing patterns that have generally favored residential over business users, rural over urban users, local over toll users. To hold the line on local prices while also making investments that, inter alia, extended residential telephone penetration from 37 percent of homes in 1940 to 94 percent in 1974 (see Fig. 1–5a), both regulator and regulated therefore put the burden on intrastate toll prices until interstate revenues began to provide relief through the allocation of increasing proportions of local facilities to interstate services (Fig. 1–27).

Federal regulators felt no pain from their constituencies so long as interstate rates continued to decrease absolutely or even relatively (see Fig. 1–8d); but just who is relieved in the states remains controversial (Fig. 1–30). Jurisdictional separations thus play a significant role in setting the relative sizes of the federal and state revenue requirement pots to be filled through pricing of all the particular interstate toll services and of all the particular intrastate toll and local services that fall in the respective pots.

In the aggregate, the pricing of the particular services in all the various pots is constraining by the overall "costs = revenue requirements" equations that regulators set when establishing allowable rates of return on total investment. Absent "predatory pricing," which all concerned agree is bad form and also unlawful, the floor price of any particular service is the cost of supplying it. But given the inherent arbitrariness of costing in any multiservice firm, that floor is foam rubber not reinforced concrete, and on it stand prices further supported through rate averaging, value-of-service, and other elastic concepts.

The Revival of Competition

Since the late sixties, the prevailing practices have come under challenges arising in significant measure from the post World War II emergence of electronics—especially computer—industries as players in an arena hitherto the exclusive domain of telephone companies both in law and in technology. Directly through orders of the Federal Communications Commission, or through jawboning by the White House Office of Telecommunications Policy, these new competitors stimulated questioning of the relationship between the costs and the prices of particular services, not unexpectedly revealing a wide variation of earnings, for given prices, among various services depending on variations of the costing method applied (Fig. 1–31).

Pricing practices might quietly have adapted to the technological changes related to the emergence of computers, as they have to earlier technological change. The upsetting factor is the revival of competition where none has existed for over 50 years, thereby creating what AT&T's board chairman described as "a degree of uncertainty with respect to the industry's future course matching that attending its birth" 100 years ago [31]. At stake are not only price changes for particular services—and the cost changes these imply for consumers—but also the implied shakeup of all existing political accommodations on costing and pricing. For the news industries, as for many other consumers, the issue sprang up in the form of possible major price increases. Attorneys for the news media voiced themselves vigorously as to this possibility, predicting the direst consequences for both the media and the public they serve.

In the next section we sketch the waves that new technologies and new competitors sent over the private line services. In Section 7 we show these waves as they wash over the news media; we also illustrate their wider spread. So long as the overall "costs = revenue requirements" equation must hold, revenues lost in one regulated service must be gained in others, and vice versa. Hence, the strong linkages among all users of all services, including every ordinary citizen. Who will gain, who will lose, how much, if what kind of competition challenges the regulated monopoly?

(Text continues on page 89 . . .)

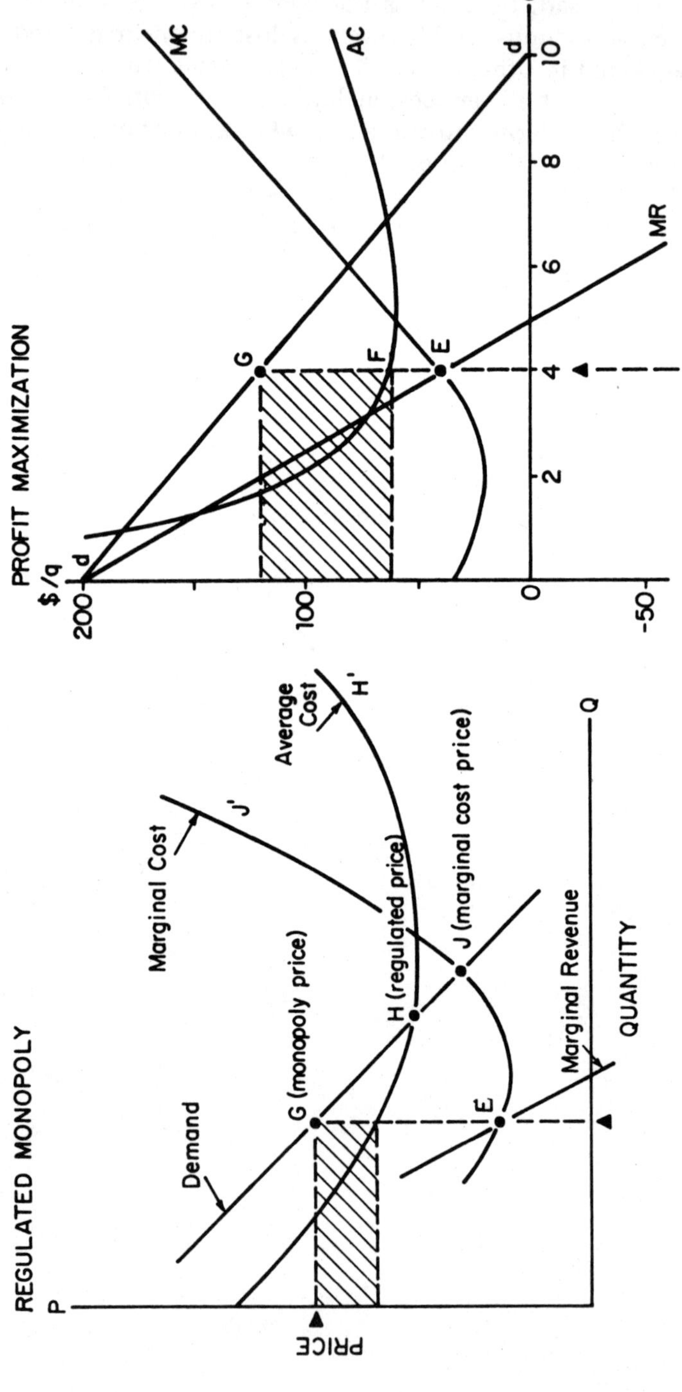
PROFIT MAXIMIZATION
$/q
200
100
0
-50
d
MC
AC
G
F
E
MR
d
q
2
4
6
8
10
(b)
REGULATED MONOPOLY
P
PRICE
Demand
G (monopoly price)
H (regulated price)
J (marginal cost price)
Marginal Cost
J'
Average Cost
H'
E
Marginal Revenue
QUANTITY
Q
(a)

Figure 1–19. Some Key Principles of Economic Theory

One of the goals of economic theory is to account for relations among demand, cost, and price. The two illustrations shown here are arbitrary, not meant to depict any specific telecommunications service.

(a) According to economic theory, total profit (shaded rectangle) is maximized in a competitive market at an equilibrium where the price G corresponds to the quantity of goods for which marginal cost equals marginal revenue E. In perfect competition, each of many small competitors sees marginal revenues and demand as one and the same horizontal line. Then G and E coincide, there is no excess profit, and utility as measured by the difference between the areas under the demand and marginal cost curves to the left of the line G–E is maximized.

(b) As Samuelson explains it: "Maximum-profit equilibrium for the unregulated monopolist is at G, directly above the intersection of MR and long-run MC with P above MC and with monopoly profit shown by the shaded rectangle. Public-utility regulation of the monopoly would set its price down to the H intersection of demand with long-run average cost; this wipes out excess-profit; more important, it brings price down closer to the Marginal Cost level at J, where marginal social costs and benefits are appropriately balanced. Ideally, P should be forced all the way down to MC at J, with the chronic loss being covered by permanent government (lump-sum) subsidy." Samuelson duly cautions that the desirability of moving from G to H involves "value judgments beyond technical economics," leaves aside equity considerations and assumes that dollars truly reflect social utilities and social costs. Since the cost of subsidy must itself be borne by someone, a full theoretical discussion of this case entails further complications.

Source: (a) Paul A. Samuelson, *Economics*, 9th ed. (New York, N.Y.: McGraw-Hill), 1973, p. 495.
(b) *Ibid.*, p. 501.

As of December 31, 1973	Revenues of All Reporting Companies ($ Million)	% of Total Revenues	AT&T Long-Lines Revenues ($ Million)	AT&T Long-Lines % of Category
Subscribers' Stations	11,850.3	46.2	-	-
Public Telephones	334.3	1.3	-	-
Local Private Lines	173.2	0.7	-	-
Message Telecommunications Service (MTS)	10,107.3	39.4	1,480.8	80.3
Wide Area Telecommunications Service (WATS)	985.2	3.8	170.5	9.2
Toll Private Lines	1,158.4	4.5	163.3	8.9
Directory Advertising and Sales	885.2	3.5	-	-
Other	145.9	0.6	30.1	1.6
TOTAL	25,639.8	100.0	1,844.7	100.0
Toll Revenues Attributed to Interstate and Foreign Service	7,746.3	30.2	1,826.1	99.0

(a)

	Percent of Total Revenues						
	1973	1970	1965	1960	1955	1950	1945
Subscribers' Stations	46.2	47.3	50.9	53.5	54.3	47.6	49.6
Public Telephones	1.3	1.6	2.3	2.6	3.1	2.5	4.1
Local Private Lines	0.7	0.8	.7	1.2	0.9	0.6	0.7
Message Telecommunications Service (MTS)	39.4	38.1	35.5	33.8	34.1	43.9	40.2
Wide Area Telecommunications Service (WATS)	3.8	2.8	1.8	-	-	-	-
Toll Private Lines	4.5	5.0	4.1	3.6	2.6	1.7	2.5
Directory Advertising and Sales	3.5	3.9	4.3	4.6	4.4	3.3	2.5
Other	0.6	0.5	0.5	0.5	0.5	0.5	0.3

(b)

Figure 1–20. Telecommunications Service Revenues

(a) Subscribers' station revenues (47 CFR §31.500) encompass conventional local service, including residential or business rentals of both ordinary telephones and also of private branch exchanges (PBX's); they also include local revenues from mileage charges, whether at flat or measured rates.

The Message Telecommunications Service (MTS) encompasses the familiar long distance call. Wide Area Telecommunications Service (WATS) is most widely known as the basis for "free" calls to the Area Code 800 numbers ("inward" WATS) supplied by airlines, hotel chains, car rental companies, and others eager to attract reservations from faraway customers. The subscribing business pays WATS charges related to the area served, but "without regard to the number of communications" (47 CFR §31.511). "Outward" WATS enables calling anyone within a specified area.

(b) Proportional revenues from the several services.

Source: Adapted from Federal Communications Commission, *Statistics of Communications Common Carriers*, December 31, 1973, Table No. 16, pp. 30 and 35; and previous editions.

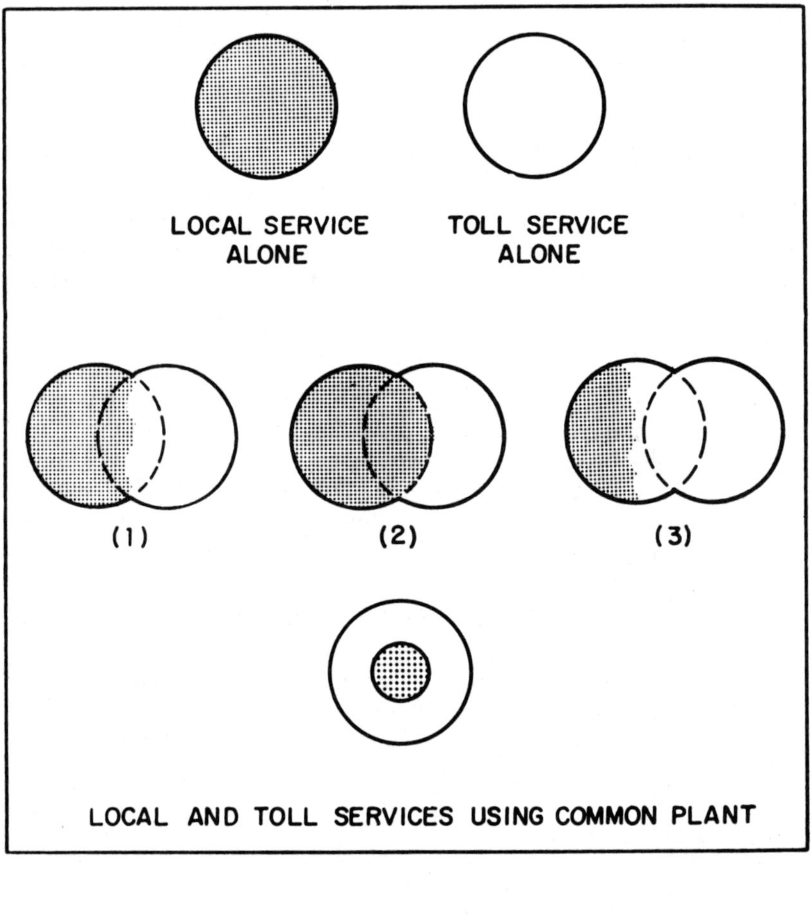
LOCAL SERVICE ALONE
TOLL SERVICE ALONE
(1)
(2)
(3)
LOCAL AND TOLL SERVICES USING COMMON PLANT

Figure 1–21. Idealized Examples of Common Cost Allocations

Shaded areas indicate local costs. Purely hypothetical costs of each service alone are shown in (a). In (b), the overlaps represent common costs, and the other areas represent costs attributable to each service. The dashed lines indicate that what is attributable and what is common is far from clear cut. In a 1975 rate case, for example, methods advanced by the U.S. Postal Service attributed 45.1 percent of total postal costs to specific services; those adopted by the Postal Rate Commission attributed 52.5 percent. The difference of 7.4 percent amounts to about 900 million dollars out of 12.3 billion dollar total annual costs.[a] Percentage swings in the telephone industry would apply to nearly 30 billion dollars. In (b1), the two services share "fully distributed" common costs in some arbitrary proportion. In (b2), toll services can be visualized as added on to local services to justify their carrying only "incremental" costs. In (b3), toll services "cross-subsidize" local services, since these bear less than their attributable costs. Some make a case for the relationship shown in (c), since nearly all local facilities are used jointly for local and toll service. The symmetry of the incremental cost argument is evident in (c).

If, as in (b1), common costs are viewed as "allocated" to the services, it can be argued that cross-subsidy or predatory pricing begins whenever costs are less than "fully allocated" or prices are less than fully allocated costs, and not merely when such relationships obtain for attributable costs. The telephone companies and, more recently, the Postal Service, prefer to visualize a pool of common costs to which services may or may not contribute, in proportions determined by *pricing* policy, thus narrowing the concepts of cross-subsidy or predatory pricing to the case of attributable costs. Still more complex conceptions of fair cost burdens are set forth in the economic literature and in regulatory proceedings.

[a] U.S. Postal Rate Commission, *Postal Rate and Fee Increases, 1973*, Vol. 1, Docket No. R74–1, (Washington, D.C.) August 28, 1975, p. 6, 14.

Figure 1–22. How Is the Mail Paid For?

. . . the concepts outlined in Figure 1–21 are illustrated here with data on the Postal Service . . .

Controversies about postal rates are mainly about how fairly service benefits and price burdens are distributed over senders and recipients of the different types of mail service.

Overall, the U.S. Postal Service meets its costs through revenues from its users plus some direct congressional appropriations. In postal accounting, attributable costs are those that vary with the volume of a service or are otherwise a fairly clear consequence of providing that service. Attributable costs for 1973 were 4,903 million dollars. Nonattributable "overhead" costs were 4,937 million dollars, making the total costs—hence the required revenues—9,840 million dollars, or 200.7 percent of the attributable costs.

Attributing attributable costs is partly a technical matter, but assigning overhead burdens is wholly arbitrary. If proportionality of overhead costs to attributable costs is taken as the yardstick of fairness—itself a controversial policy choice—then, at the extremes, government and priority mail are seen to carry disproportionately high shares of overhead costs and publishers and library mail disproportionately low shares. (Data from *Statistical Abstract of the United States*, 1974, p. 497.)

MAIL SERVICE	ATTRIBUTABLE COSTS IN $ MILLIONS	REVENUES ATTRIBUTABLE + OVERHEAD COSTS) AS PERCENT OF ATTRIBUTABLE COSTS
1973 Total	4,903.1	200.7
First-class mail	2,414.2	189.6
Airmail	189.0	112.6
Priority mail	150.7	233.6
Second-class mail	362.2	51.4
Publishers mail	356.7	50.7
Outside the county:		
Regular rate publications	257.8	57.4
Other	68.4	27.2
Within the county and fees	30.4	46.7
Controlled circulation publications	20.5	179.5
Third-class mail	759.6	153.0
Bulk rate	624.3	159.3
Single piece rate and fees	135.3	124.0
Fourth-class mail	526.8	144.1
Zone rate mail	401.3	160.0
Parcels	386.0	158.6
Catalogs and fees	15.3	194.8
Special fourth class rate	116.0	97.2
Library rate and fees	9.5	43.2
Government mail	98.2	411.4
International mail	187.2	169.1
Special services	192.7	166.9
Other	2.2	--

1	American Telephone and Telegraph Company
	(including Bell Laboratories and Western Electric)
25	Bell Operating Companies
1	Western Union
ca. 1600	Independent telephone companies
5+	Specialized common carriers and domestic satellite carriers
X_1	Value-added service companies
X_2	Teleprocessing service companies
X_3	Electronics and business machines companies
X_4	Trade associations
	(Association of Data Processing Service Organizations, Inc., Computer and Business Equipment Manufacturers Association, Electronic Industries Association, National Telephone Cooperatives Association, North American Telephone Association, U.S. Independent Telephone Association, etc.)
1	Congress of the United States
1	Federal Communications Commission
1	United States Postal Service
X_5	Other federal executive agencies
	(Internal Revenue Service, Rural Electrification Administration, Rural Telephone Bank, White House Office of Telecommunications Policy, etc.)
X_6	Courts
50	States
	(regulators, legislators, consumer agencies, etc.)
1	National Association of Regulatory Utility Commissioners
X_8	Municipalities
	(local taxation, pole rights and other rights-of-way, etc.)
X_9	Investors supplying debt and equity capital
X_{10}	Consumers and consumer groups
	(including individuals, corporations, consumer groups, and government)
X_{11}	Foreign nations

Figure 1–23. The Telecommunications Players

The principal players in the domestic telecommunications arena are many, varied, and sometimes at cross-purposes, sometimes in coalition.

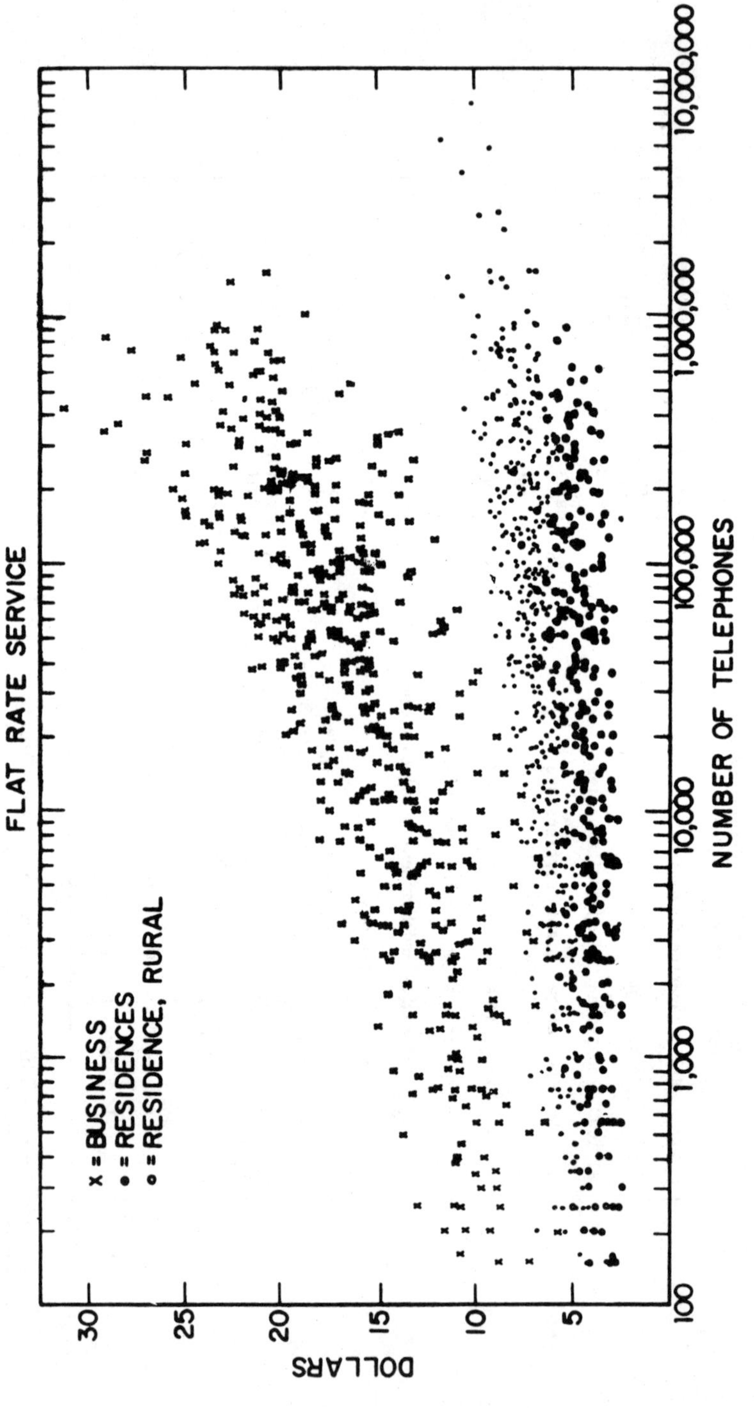
FLAT RATE SERVICE
x = BUSINESS
• = RESIDENCES
◦ = RESIDENCE, RURAL
DOLLARS
30
25
20
15
10
5
100
1,000
10,000
100,000
1,000,000
10,000,000
NUMBER OF TELEPHONES

(a)

Figure 1–24. How Telephone Service Is Paid For

Monthly rates are shown for basic flat rate main station service for individual (one-party) business and residence telephones and for rural service that extends into the hinterlands with up to eight parties on a line but no charge for mileage.

Rates vary with the number of telephones in the subscriber's service area, i.e., the number of phones that may be reached without additional charges. Through price averaging, the variation within any one service may or may not reflect cost variation (Fig. 1–12c). The concept of "value-of-service" pricing is most clearly illustrated by the large differences between the prices and the variances for business and residential services that use essentially the same facilities. In some of the largest service areas only message rate service is offered, especially for businesses.

The data show rates for Bell Companies. Current data are not available for most independent companies, since the U.S. Independent Telephone Association stopped compiling such figures ca. 1970. Data for California, Missouri, New York, and Texas are not included.

Source: Data compiled from National Association of Regulatory Commissioners, *Exchange Service Telephone Rates in Effect June 30, 1974* (Washington, D.C.: NARUC).

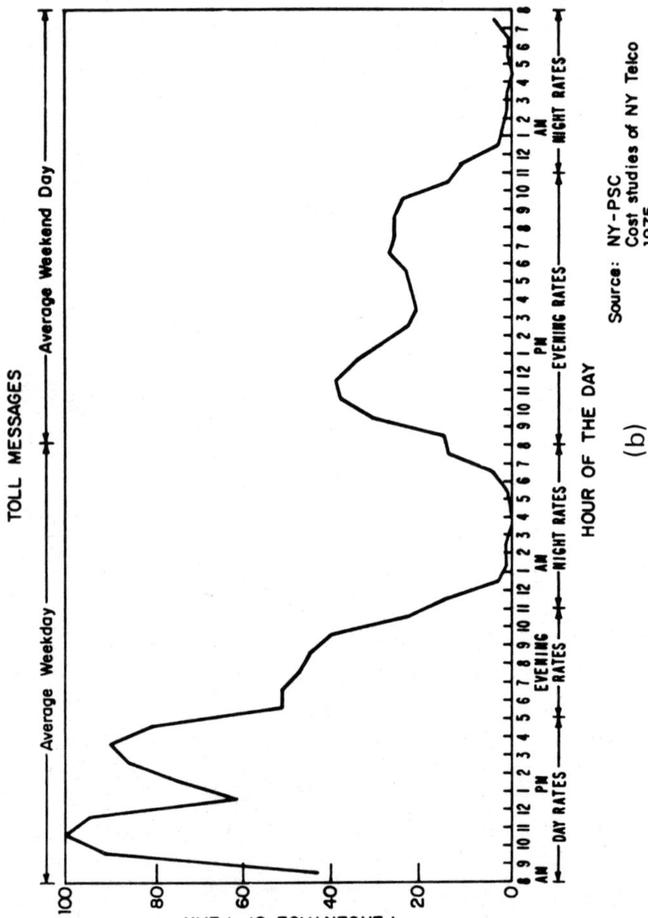

Figure 1–24. cont.

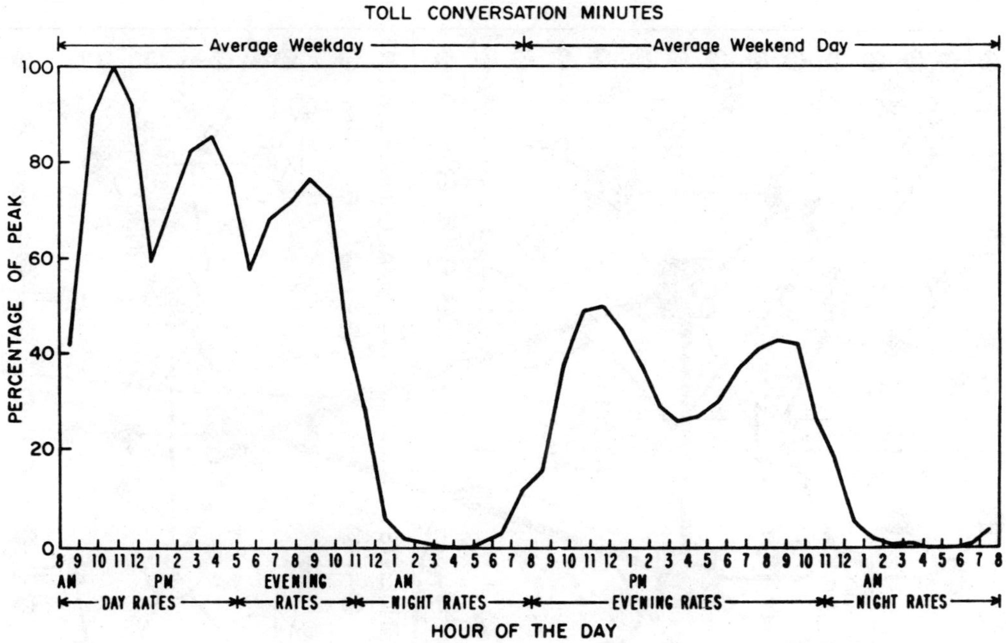

(c)

Figure 1–24. cont.

(b) and (c) Since peak toll traffic occurs during business hours, it can be argued that plant facilities must be designed to meet that peak. Lower prices for off-peak use are then justified as a fair encouragement of the use of otherwise idle facilities. Alternatively, higher peak prices are justified as a fair way to ration scarce fixed peak time resources. Note how evening conversations tend to be longer than daytime conversations.

Source: NY–PSC. Cost studies of NY Telco. 1975.

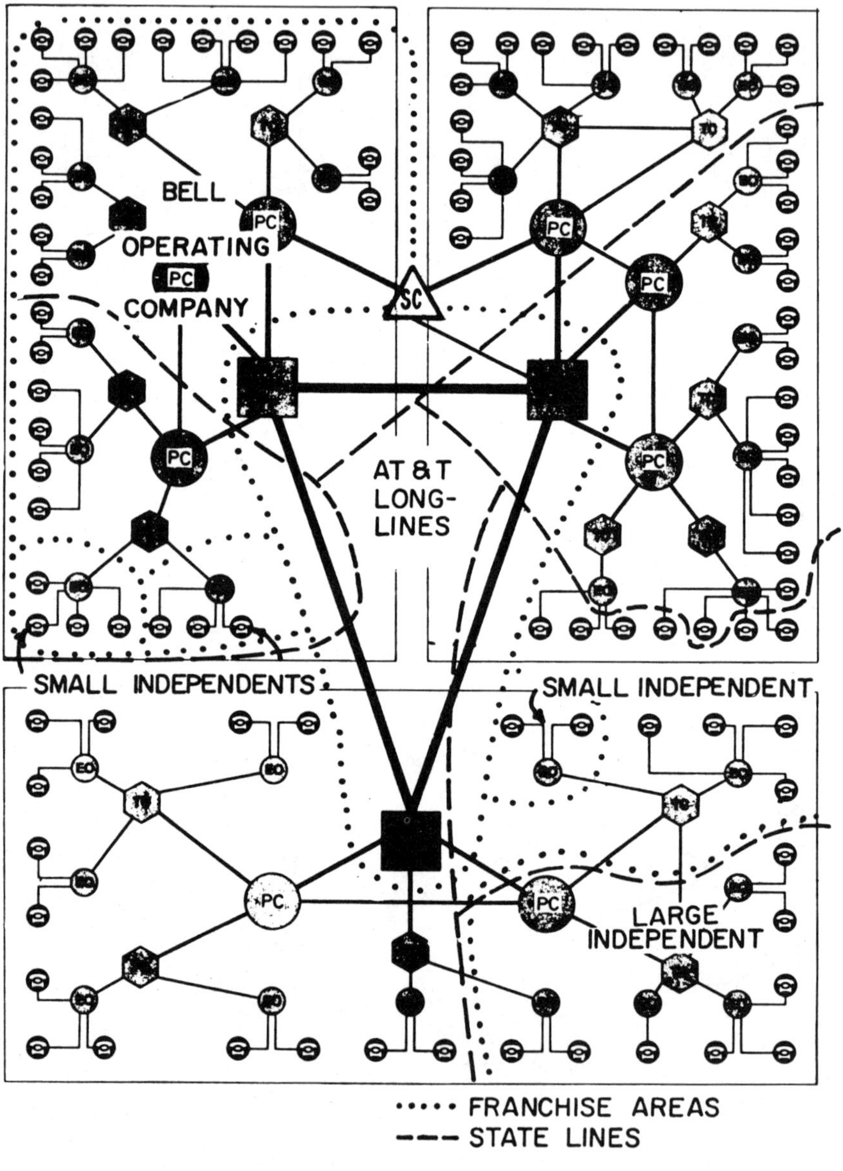
BELL
PC
OPERATING
PC
COMPANY
SC
PC
PC
PC
PC
AT & T
LONG-
LINES
SMALL INDEPENDENTS
SMALL INDEPENDENT
PC
PC
LARGE
INDEPENDENT
••••• FRANCHISE AREAS
---- STATE LINES

(a)

Figure 1—25. Telecommunications Ownership and Jurisdictional Patterns

Globally, the network of Figure 1—13 crosses numerous national boundaries under complex ownership patterns. Within the United States, AT&T, over 20 Bell operating companies and some 1,600 independent telephone companies provide service within and across the boundaries of 50 states. Overlapping ownership and jurisdictional lines are shown schematically in (a). The indicated Long Lines boundary is purely symbolic, since it corresponds to no well defined geographic area, and is not limited to the highest level of the switching hierarchy. The federal presence manifests itself through the state line crossings that inject the interstate commerce clause of the U.S. Constitution into telecommunications matters. The geographic coverage of AT&T's high capacity "spine" is shown in (b). Some characteristics of overall geographic coverage are tabulated in (c). The average telephone density (Telephone/Mile^2) in Bell territories is twice the national average.

Source: (b) George P. Mandanis et al., *The Domestic Telecommunications Industry: Economic Behavior, Competition, and Public Policy*, R73—40 (San Rafael, Calif.: Systems Applications, Inc.), May 10, 1974, p. 255.

(c) B.A. Hart and R.F. Martin, *Telephone Density and Telephone Company Area Data*, OTR 73—18 (Washington, D.C.: Office of Telecommunications), 1973.

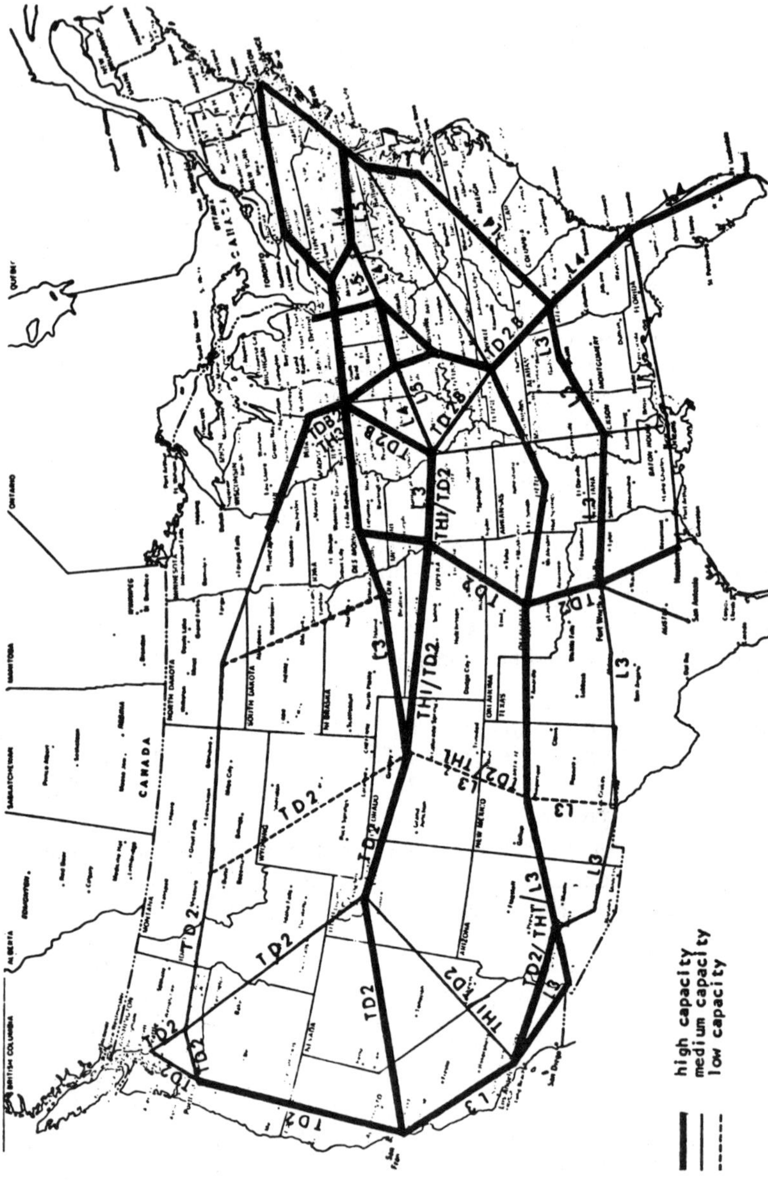

(b)

Figure 1–25. cont.

	% of Telephones in U.S.	% of Area in U.S.	Telephones/ Mile^2
Bell	83.3	41.9	82.0
General Telephone and Electronics	7.6	7.9	39.7
United Telecommunications	2.0	3.6	22.5
Other	7.2	33.1	8.9
Undesignated	0	13.5	0
U.S. Average			41.2

(c)

Fig. 1–25. cont.

All $ in Millions

	Total Revenues	Intrastate $	Intrastate %	Interstate & Foreign $	Interstate & Foreign %
Local Service	12,357.8	12,357.8	100	-	-
Toll Service	12,250.9	4,504.6	36.8	7,746.3	63.2
Both Services	24,608.7	16,862.4	68.5	7,746.3	31.5
AT%T Long Lines				1,826.1 (23.6% of Interstate and Foreign)	

(a)

	#(Billion)	%	Average #/ Phone	Average $/ Phone Call
Local Calls	200.1	93.7	1,694	.059
Toll Calls	13.4	6.3	113	.754

(b)

	# of Phones (Million)	%	Revenue Share
Business	31.7	26.3	53.9
Coin	1.5	1.2	2.8
Residence	87.3	72.4	43.3
Total	120.5	99.9*	100.0

*Error due to rounding off.

(c)

Figure 1–26. Jurisdictional Shares of Revenues

Jurisdictional shares of revenue are estimated in (a) from data in Figure 1–20. Though there are 16 times more local calls than toll calls (b), the average toll call brings in about 13 times the revenue of the average local call. The higher basic business rates (Fig. 1–24), the lower number of business phones (c), and different business and residential use of additional ("vertical") services and their calling habits are reflected in the resultant revenue shares.

Source: Adapted from Federal Communications Commission, *Statistics of Communications Common Carriers*, December 31, 1973, Table No. 16, p. 30, 32, and 35; and from American Telephone and Telegraph Company, personal communication.

	1973 Total Investment ($ Millions)	Interstate Allocation (% of Total Investment in category as allocated to interstate costs according to successive separations plans)							1973 Inter-state Plant Investment ($ Millions)	Annual Carrying Charge (Fraction of Investment)	1973 Interstate Revenue Requirements Associated with Plant ($ Million)
		1947	Charleston 1952-57	1957	Denver 1966	FCC-16258 1967	FCC-17975 1969	Ozark 1971			
Station Equipment	322.7	6.91	11.66	12.11	10.41	18.79	21.50	23.96	77.3	.30	23.2
Exchange Circuit Equip.	43.3	4.48	6.89	12.35	7.05	11.25	12.61	15.15	6.6	.25	1.6
Toll Switching Equip.	3.1	6.91	6.89	37.46	37.46	37.46	37.46	37.46	1.2	.25	.3
Local Dial Switching Equip.	258.2	6.91	6.89	6.89	9.47	9.47	9.47	15.43	39.8	.25	10.0
Exchange Outside Plant	545.2	6.83	11.66	11.82	10.04	18.11	20.71	24.09	131.3	.25	32.8
Interexchange Outside Plant	33.1	42.67	42.67	42.67	42.67	42.67	42.67	42.67	14.1	.25	3.5
Interexchange Circuit Equip.	119.8	47.91	47.91	47.91	47.91	47.91	47.91	47.91	57.4	.25	14.4
	1,325.4							*24.72	327.7	*.26	85.8

* Weighted Average

Figure 1–27. United Telecommunications—Allocation of Local Facilities Costs to Interstate Revenue Requirements

The percentage of local plant attributed to interstate operations hence interstate revenue requirements rose sharply over successive separations plans (see also Fig. 1–30).

Source: United System Service, Inc., Response to FCC Docket No. 20003, April 1975, Exhibit 4.

YEAR	Long Lines	Operating Companies	Ratio
1950	$59	$114	1.9
1952	56	114	2.0
1954	43	111	2.6
1956	39	109	2.8
1958	38	112	2.9
1960	32	106	3.3
1961	31	104	3.4
1962	29	101	3.5
1963	27	97	3.6

Figure 1–28. Relative Costs of Interstate and Intrastate Facilities

Between 1950 and 1963, AT&T Long Lines' book costs per equivalent circuit mile were halved while those of operating companies dropped by only 15 percent of their original value. This reflects the fact that Long Lines carries heavily concentrated traffic that can benefit from the economies of scale illustrated in Figure 1–8. State toll services handled by operating companies operate with transmission systems at the lower end of the capacity range. For fixed traffic densities, a single long haul link will cost less per mile than several short haul links adding up to the same length, since total termination costs (multiplexers, etc.) are lower for the former than for the latter. These differences are reflected in intrastate toll rates higher than interstate and in pressures toward increasing allocations of costs through separations to the interstate services.

Source: Richard Gable, *Development of Separations Principles in the Telephone Industry*, (East Lansing, Mich.: Institute of Public Utilities, Michigan State University), 1967, Table XVIII, p. 146.

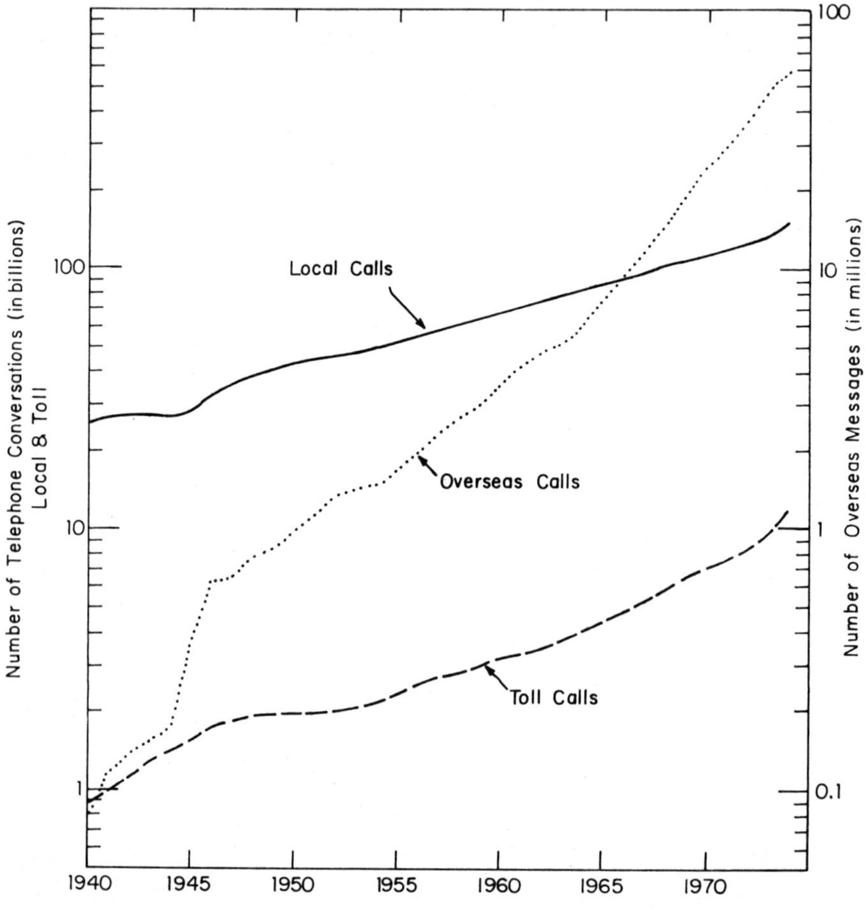

Figure 1–29. Relative Demand for Local and Toll Calls

Source: American Telephone and Telegraph Company, *Bell System Statistical Manual*, §800, Volume of Communications, p. 803 and 807.

Date	Nature of Separations Change	Net Decrease in Intrastate Revenue Requirements ($ Millions)	Reported Decrease in Intrastate Rates ($ Millions)
1942	Introduction of S-S Allocations	22.0	3.9
1947	Amendments with 1947 Manual	19.0	-0-
1951	New Toll Coefficients	17.0	-0-
1951	Charleston Modifications	32.8	-0-
1956	Modified Phoenix Plan	40.0	-0-
1960	New Toll Coefficients	4.0	-0-
1962	Reclassification of OSP AC/S; measurement of minutes of use; local comml ops.; etc.	46.0	40.0
1964	Revenue Acct. Modifications	.9	-0-
1965	Denver Plan	93.5	16.9
1970	Ozark Plan	130.0	?

Figure 1–30. Relief Through Separations

Successive separations plans shifted intrastate revenue requirements to interstate. Depending on where the commentators sit, the resulting impact on interstate prices is interpreted either as evidence of additional profits for the local telephone companies or as a sign of prices generally holding firm in spite of increased costs and lower rates of return incurred by these local companies (see also Fig. 1–27).

Source: 1942–1965: Richard Gabel, *Development of Separations Principles in the Telephone Industry* (East Lansing, Mich.: Institute of Public Utilities, Michigan State University), 1967, Table X, p. 128.

1970: 26 FCC 2d 260, Docket No. 18866.

[Bell System interstate services—12-month period ended July 31, 1971]

Category	Percent						
	Method 1	Method 2	Method 3	Method 4	Method 5	Method 6	Method 7
Message telecommunications service	8.6	7.9	7.0	7.7	7.7	7.2	7.3
Wide area telecommunications service	9.4	13.8	17.9	15.4	14.9	17.1	16.7
TELPAK	5.4	6.2	11.0	6.9	7.7	10.3	9.6
Private line telephone	4.0	5.0	8.1	5.2	5.0	5.7	5.3
Private line telegraph	5.3	6.3	7.2	6.4	6.4	6.6	5.7
Audio/radio	3.0	5.7	13.9	6.0	5.5	6.4	4.1
Television	4.9	5.1	6.1	5.8	5.8	9.2	8.3
Other[1]	3.1	2.7	4.4	2.8	2.6	3.2	3.3
Total	7.8	7.8	7.8	7.8	7.8	7.8	7.8

[1] Includes TWX results, prior to the sale of that service to Western Union effective Apr. 1, 1971 (see PF 224).

Figure 1–31. Price/Cost Relationships Under Variants of a Costing Method

Ratios of net operating earnings [(price × demand) – current costs] to net investment (cost of facilities) for Bell interstate services vary depending on the costing method. Earnings on television services, for example, range from a low of 4.9 percent under Method 1 to a high of 9.2 percent under Method 6. All seven methods are of the "fully distributed cost" type, with a share of common costs assigned to each service. In this case, their application to AT&T is favored by competitors, opposed by AT&T. Numerous studies have been made, with different types of methods, and different outcomes and interpretations. Another type of methods, "incremental costing," is discussed in Section 7.

Source: U.S. Congress, Senate, Committee on the Judiciary, Subcommittee on Antitrust and Monopoly, Hearings on S. 1167, Part 2: The Communications Industry, July 30 and 31; August 1 and 2, 1973; p. 1626.

6. THE EMERGENCE OF COMPUNICATIONS AND TELEPROCESSING: NEW COMBATANTS ON THE FIELD

Fascinating as is the brawling history of early telephony, we trace only in barest terms the antecedents of the price and service warfare that smoldered for 50 years, then flared up once again following the post World War II advent of consumerism, of electronics in general, and of computers in particular.

By 1921, the current pattern of coexistence of the Bell System and of the independent telephone companies as geographic monopolies, and also the nationwide interconnection of telephone companies, had been guaranteed by AT&T's Kingsbury Commitment of 1913 and the enactment of the Willis-Graham Act of 1921. After 1926–27, radio, television and telephony went their distinct but linked ways; and, after World War II, Western Union's telegraph service went into a coma [32]. The telecommunications battlefield, however, was fired up by a boom in electronics, especially in computers, ironically boosted by the invention of the transistor at Bell Laboratories in 1947. The news media, like all other consumers depending on a telecommunications backbone, were caught in the resulting crossfire.

Compunications Science and Technology; Telecommunications and Teleprocessing Functions

A major underlying technological factor is the merger of computer and communications technologies. This technological convergence has thrust hitherto clearly distinct telephone, business machine electronics, and other industries together in one battlefield, with old and new customers as the prize.

The classic block diagram for a computer (Fig. 1–32) seems to differ profoundly from Shannon's model (see Fig. 1–11) of a communications channel. But "input" and "output" in the one and "transmitter" and "receiver" in the other are only different jargon. The heart of the matter seems to be the difference between the groupings of processor, storage and control in the one and of channel and noise source in the other. Shannon divorced the source from the transmitter and the receiver from the destination in order to differentiate clearly between the message originating at a source (e.g., an author's thoughts) and the encoded signals transmitting that message over a channel (e.g., words on paper). The same distinction applies to computer systems. Explicitly replacing each line in Figure 1–32 as indicated for the input-processor link, makes it plain that computer systems subsume communication systems.

We do not normally see a hand-held electronic calculator or a microcomputer [33] as a communications system. Usually that viewpoint is saved for distributed computer systems, where an input terminal might be in New York, the storage and the processor in Cleveland, and an output terminal in Los Angeles or London. Yet the microcomputer designer, who measures time in nano- or pico-seconds (10^{-9} or 10^{-12} sec.) is very conscious of microscopic transmission paths. Just as the lines in Figure 1–32 hide communications systems, so do the channel, transmitter and receiver in Figure 1–11 hide computers, and so communication systems subsume computer systems. That transmission and reception subsume storage and processing is implicit, for example, in Shannon's analysis of encoding (a form of processing) and of the "delay . . . generally required to approach the ideal encoding." Such delay [34] must be realized by some form of storage.

By 1948, storage, processing, and control had long been commonplace in telegraphic switching, even if not explicitly labelled as such. An incoming telegram was punched into a tape. The switching operator would take the tape from the receiver, read the destination codes, and feed the tape into a sender connected toward the intended destination. The torn tape was the storage medium and remains so in older systems; the processor was—and in many systems still is—the human operator, who also controlled the process. Such functions persist, albeit in automated form, as so-called "store-and-forward" or "message" switching in many compunications systems.

In the earliest telephone exchanges, the storage function was not so plain to see. The operator remembered where to route an incipient conversation long enough to do "line" switching by sticking the right plug into the right hole for making a wire path from transmitter to receiver. The message itself was stored in the system only when a nosey operator eavesdropped. Processing took place when the sound waves were encoded into electrical signals at the mouthpiece. However, these computer functions became more explicit as switching automation progressed. By 1906, the Bell System had introduced the "sender," a device "that received the telephone user's dial pulses, stored them, and controlled the setting up of the call via the office switch train" [35]. The modern electronic telephone switch is explicitly a computer.

The abstract equivalence between basic computer and communications is made ever more concrete as differences disappear between the "stuffs" that flow in computer and in communications facilities and as the distinction between message and line switching facilities vanishes. The discrete on-off zero-one dot-dash digital signals of telegraphy and of computers are akin to one another but distinct from the traditional continuously varying analog signals of telephony. Nowadays, however, digital pulse code modulation (PCM) combined with time division multiplexing (TDM) [36] is gradually becoming the preferred form of encoding transmission even for originally analog voice messages (Fig. 1–33).

Computers developed by Bell and by other communications and computer manufacturers in the United States or abroad switch digital signals by assigning them to some appropriate bandwidth or time slot on an outgoing trunk without setting up the solid "copper paths" hitherto necessary to route even digitally encoded signals. The distinction between line switching and message switching functions thus ceases to reflect sharp organic differences between the facilities for doing them. The "channel" function also is no longer identical to a solid wire facility, even if the channel serves as a private line. In a PCM/TDM system, the "path" becomes thin slices of time available in varying frequency bands over several transmission links realized in diverse technologies. In each link this path is defined by the multiplexers that bundle a signal with many other signals for carriage over the same physical facility. Depending on the delays in assigning messages to time slices, similar facilities will be seen as performing either line switching or message switching functions.

The current flowering of both PCM transmission and computers stems from their common concrete realization in the low cost technology of large scale integrated (LSI) solid state circuits [37]. In this medium, the abstract identity of computers and communications become fully concrete. In an earlier work [38] one of us reflected this merger by coining the descriptive term "compunications" for the combined science and technology. No one seriously quarrels over the reality of compunications *science* and *technology*. The battles are over *functions*. "Telecommunications" is still widely used to denote the function of communicating over distance. "Teleprocessing" is becoming increasingly common to describe geographically distributed data processing functions linked by telecommunications. Although the Federal Communications Commission attempted to settle the matter by launching an inquiry in 1966, issuing a decision in 1971, and, after court interventions, revising that decision in 1973 [39], titanic battles continue to rage over the boundary—if there is any—between the telecommunications and teleprocessing functions and industries. These two worlds, building on a common base of compunications science and technology, are as distinct or indistinct in their functions as are AT&T and IBM, "regulated" and "free" enterprises, or "common" and "private" goods or services.

This confounding of telecommunication and teleprocessing functions in systems built on a compunications base is illustrated by UPI's operations. UPI has installed a computerized system to edit, abstract, and queue up stories to be sent out on the UPI wires. Stories arriving in UPI's New York switching and computing center over their international network (see Fig. 1–18) or their domestic private lines are displayed and edited on computer-linked cathode ray tube terminals. After editing, the press of a button returns the story to a computer in that New York switching center. Then the story is abstracted and placed on a queue according to its assigned priority. This priority determines if and when the story will be sent out on which of UPI's different service wires. In an increasing number of cases, these connect directly to client newspapers' computers, not to their teletypewriters. A copy of the abstract is automatically sent to client editorial offices, and the story itself is stored for a time in a computer data bank, where a remote editor interested by the abstract may retrieve it.

UPI is a private service, at least by virtue of a specific exemption for press services written into the history of the term "common carrier" in the Communications Act of 1934 [40]. Nevertheless, UPI has always followed a conscious corporate policy of serving anyone who can pay for their service, in contrast to AP's once restrictive policies as to whom they would serve. However, since AP's restrictive service policies were declared illegal under the antitrust laws by the Supreme Court in *United States* v. *Associated Press* [41], the press service practice of serving all who can pay, just as public utilities must, is a legal mandate as much as voluntary corporate policy.

New Competitors: The Interconnectors

The merger of computer and communications technologies and functions has not, however, been limited to specialized services like UPI. The cost revolution that has produced the $5 transistor radio and the $10 calculator has also spawned numerous other products and services dependent on attaching diverse equipment to incoming and outgoing telephone lines. Among these are equipments and services such as UPI uses. But, except for or through equipment owned by the telephone company, or by specially designated communications common carriers, telephone companies' tariffs prohibited connecting so-called "foreign" attachments or networks to any but private lines until the late 1960s [42]. Computer terminal users in the 1960s attempted to forestall charges of tariff violation by hooking terminals up to acoustic couplers that beeped tones into ordinary telephone handsets.

Such an acoustic/inductive device produced by Carter Electronics Corporation for linking the base station of a private mobile radio system to the common carrier network provided a test case. After preliminary legal jockeying, a unanimous FCC held in June 1968 that the prohibitions of "foreign" attachments "are, and have since their inception been, unreasonable, unlawful, and unreasonably discriminatory under the Communications Act of 1934" [43]. Noting that "the vice of the present tariff . . . is that it prohibits the use of harmless, as well as harmful devices," the Commission authorized interconnection so long as it "does not adversely affect the telephone company's operations or the telephone company's utility to others."

That principle was easier to state than it is to apply. As might be expected, the battle shifted to the definition of "harm." The question of "technical harm" emerged as one of standards and costs, for, as found in 1970 by a panel of the Computer Science and Engineering Board of the National Academy of Sciences, even though "uncontrolled interconnection can cause harm to personnel, network performance, and property," the nature of the harm and protection criteria "can be specified explicitly enough to be understood and acted upon properly by people with normal technical competencies" under suitably defined circumstances [44]. Debates over "economic harm" are still raging, as will be illustrated in Section 7.

New Competitors: The Specialized Common Carriers

The number of "foreign" attachments has grown since the Carterfone decision (Fig. 1–34), but on another front, the battle over other "adverse effects" continues. The interconnection of foreign attachments can be seen as affecting only the outer reaches of the telephone network. The emergence of so-called specialized common carriers (SCC), however, is seen by the telephone companies as striking at the economic and political heartland. (However, to the extent that terminals are "smart" computers and are themselves connected by telecommunications to other terminals, any inside/outside distinction reflects a vanishing difference.)

In Sections 3 and 4 we noted the "classic" purposes for which television and other news services want private line services. The desire for unfailingly high transmission quality is even more intense when digital data are transmitted among computers and computer terminals. Compensating for errors increases computer costs. Uncompensated interruptions can mean the loss of network synchronization and the garbling of signals which require, at the least, retransmissions or other measures to avoid an unintelligible telegram, an erroneous report on a suspected criminal, or a misguided rocket. The need for high quality becomes increasingly important given the absence, in many data transmission systems, of human intervention on reception of the bit stream to "get the message from the context" or to "hear" or "see" the correct message even if there are actual transmission errors [45].

The rise of specialized common carriers is attributed in part to growing needs for data transmission services (Fig. 1–35). In August 1969 the FCC approved an application by Microwave Communications, Inc., to offer private lines through "a limited common carrier microwave radio service, designed to meet the interoffice and interplant communications needs of small businesses" [46], needs with unique and specialized characteristics [47]. By mid 1975, the specialization of the SCCs was questionable, as MCI and Southern Pacific Communications, another SCC, were carrying mainly private line voice traffic. Other SCCs, like Datran, indeed specialized, at least initially, in digital data services.

Whatever the stated purpose and the emerging reality, SCC competition over private line services has made waves in the costing and pricing accommodations described in the preceding section. How this happened is detailed in the next section. The waves are spreading. The FCC decision approving the original SCC application noted that "the proposed services would be limited to transmissions between MCI's microwave sites, making it incumbent upon each subscriber to supply his own communications link between MCI's sites and his place of business" (i.e., his own loop service) [48]. This meant, of course, that the local loops would be leased from the classical common carriers. Indeed, SCCs or their clients have demanded access also to Bell CCSA service so as to provide switching among unswitched SCC lines, and to Bell FX service, tied via SCC lines to their customers' premises. The resulting new facilities are indicated very schematically in Figure 1–36. Actual routes as of early 1975 are shown in Figure 1–37a. Comparison with Figure 1–25b shows the coincidence between these new routes and some of AT&T's busiest routes [49], all of them links among major business centers.

The worldwide growth of computer and electronics industries has enabled Datran to realize its switched digital network design by drawing on such resources as Nippon Electric Company and Burroughs Corporation, as well as on more classical suppliers of the independent telephone industry such as Stromberg-Carlson. Such non-Bell support is needed because, under the terms of a 1956 antitrust consent decree, Western Electric, Bell's manufacturing subsidiary, is enjoined from selling "any equipment which is of a type not sold . . . to companies of the Bell System" for ordinary common carrier services [50]. So long as the Bell System was the dominant source of compunications know-how, one side effect of this pro-competitive prohibition was to make it technologically difficult for competition over network facilities to arise. But the merger of computer and communications technologies means that Bell no longer has what might be considered a monopoly on compunications know-how. The entry of other players into the communications manufacturing market might affect prices and revenues of Western Electric, thereby sending yet another wave over the tossing telecommunications price setting processes.

Other New Competitors

Other new or proposed developments are sending yet more waves over the traditional structure of the telecommunications system. Over and above earthbound activities, the development of communications satellites—deserving a detailed survey of its own—has spurred both IBM (through Satellite Business Systems, Inc., a subsidiary shared with Comsat General Corporation and The Aetna Casualty and Surety Company) and the television networks to propose entry into the transmission field, perhaps to turn both of them from large scale users into major suppliers. In 1975, RCA began supplying satellite transmission facilities to Home Box Office, Inc., a Time-Life Inc. subsidiary, to link up CATV systems into pay-TV networks.

Besides the various organizations who are building or contemplating building physical transmission facilities, there are the so-called "packet" or "value-added" networks (Fig. 1–38)—AT&T calls them Composite Data Service Vendors—offering data related services at the border between telecommunications and teleprocessing. Still less tangible are brokerage services that lease transmission facilities wholesale; using widely available compunications devices, they, or their customers, subdivide and retail this capacity, leading to still further controversies [51]. Numerous unregulated teleprocessors (Fig. 1–39) are gaining experience with telecommunications techniques. Cable television is seen by some as a source of potential competition for the local telephone company in the market for local information distribution. And, further on in the future, the development of optical means of carrying and processing information may affect the structure of compunications systems and industries in ways at which we can only guess. As the established carriers responded to these competitive pressures, attorneys for the news media in turn reacted vigorously.

(Text continues on page 108 . . .)

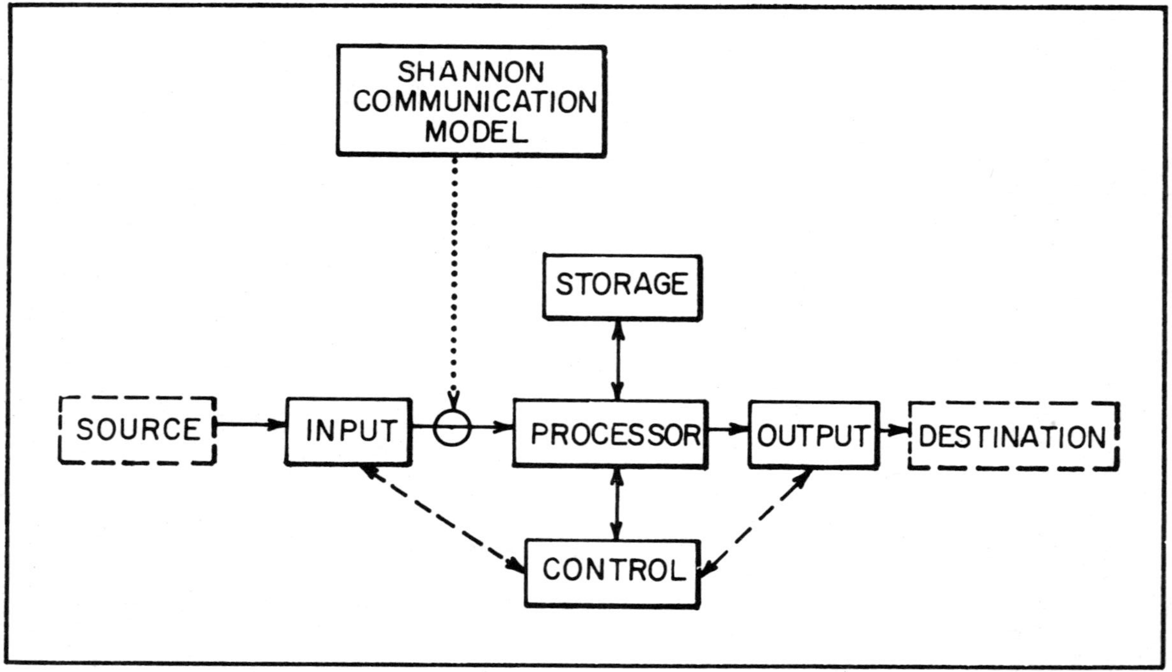

Figure 1–32. The Merger of Computers and Communications

The classic view of a computer identifies input, output, storage, processing and, especially, automatic control as its distinctive functions. However, distributed, the physical devices doing those functions are linked up by communications channels. A computer doing useful work also accepts messages from a source and delivers messages to some destination, just as communications channels do.

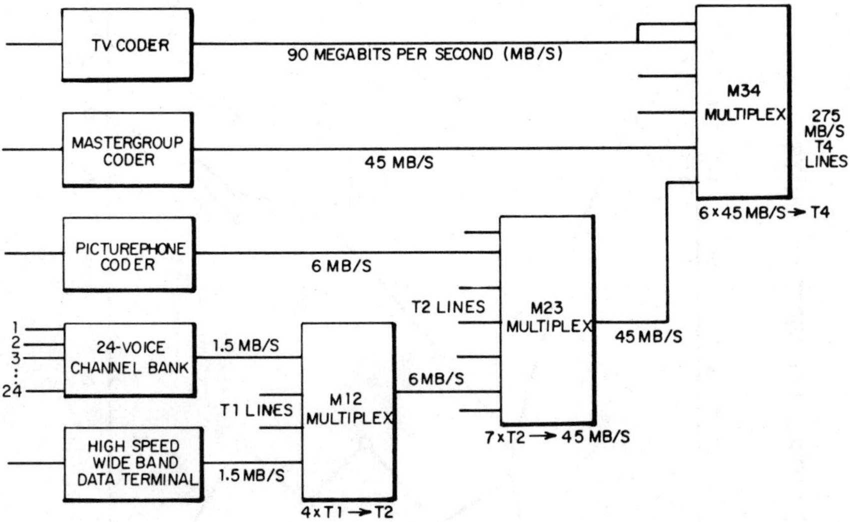

Figure 1–33. Digital Transmission of Analog and Digital Messages

Digital encoding terminals are being developed to provide time-division multiplexing in a still evolving hierarchy. All incoming signals, unless they are already in digital form, will first be translated by pulse code modulation into strings of uniform pulses and spaces representing bits, or binary digits (0 and 1). The bits representing several signals are interleaved to produce a composite signal that is unscrambled at the end of its journey. Thus 24 voice channels can be combined by a "channel bank" to produce a signal of 1.5 megabits (1.5 million bits) per second, which can be carried by a T1 wire pair. Four T1 signals can be multiplexed into a T2 signal of six megabits per second and fed into a M23 multiplexer capable of combining seven such signals into one signal of about 45 megabits per second. Each six-megabit T1 pulse stream is capable of carrying 96 voice signals or one "Picturephone" signal. The M34 multiplexer will be able to combine three television signals, or their equivalent, into one signal of 275 megabits per second. Because of the addition of extra "housekeeping" bits, the composite signals exceed simple multiples of the input signals.

Source: Ernest R. Kretzmer, "Communication Terminals," *Scientific American*, Vol. 227, No. 3, September 1972, p. 135.

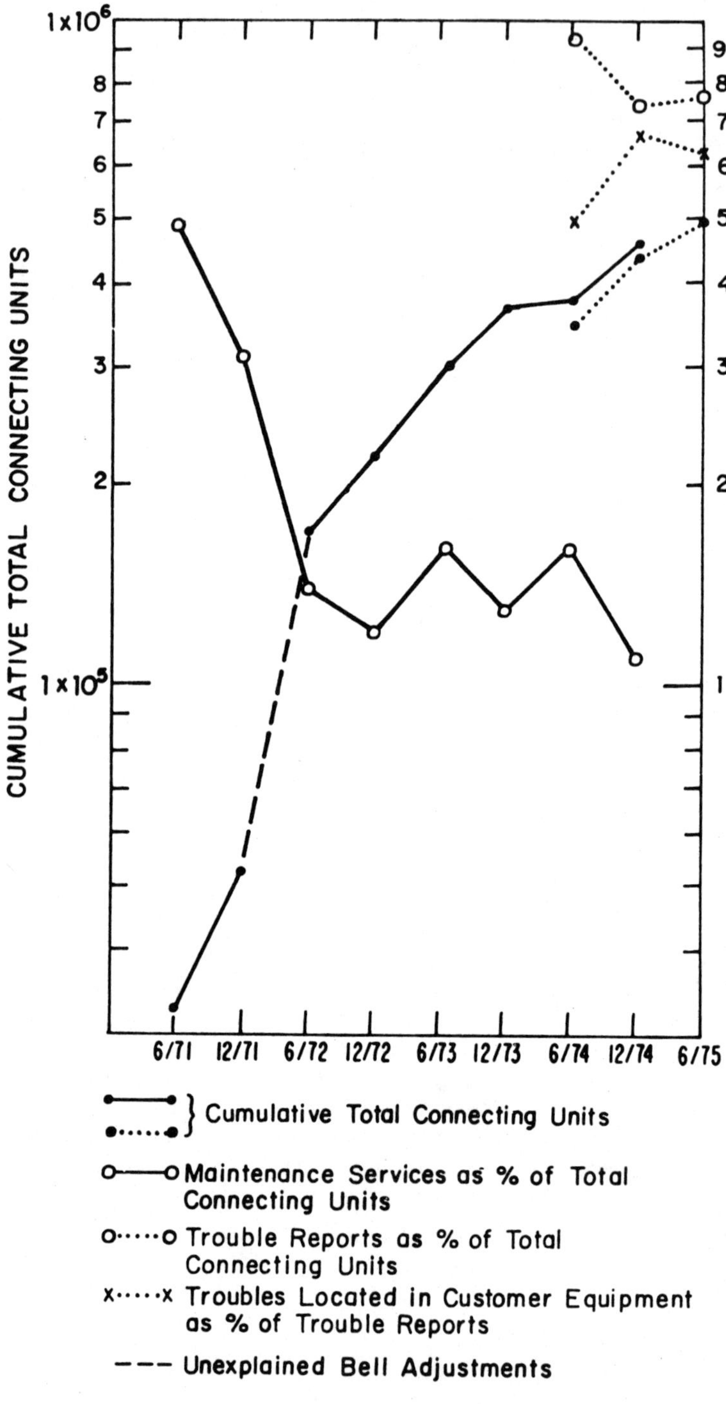
1x10^6
8
7
6
5
4
3
2
1x10^5
CUMULATIVE TOTAL CONNECTING UNITS
9
8
7
6
5
4
3
2
1
PERCENT
6/71 12/71 6/72 12/72 6/73 12/73 6/74 12/74 6/75
Cumulative Total Connecting Units
Maintenance Services as % of Total Connecting Units
Trouble Reports as % of Total Connecting Units
Troubles Located in Customer Equipment as % of Trouble Reports
Unexplained Bell Adjustments

Figure 1–34. Growth of "Foreign" Attachments to the Telecommunications Network

The number of Bell-installed connecting units linking "foreign" attachments to the telephone network has grown rapidly since the Carterfone decision. The definition and interpretation of resulting "troubles" is still in flux.

Sources: 6/71–12/72: Dittberner Associates, Inc. (Washington, D.C.). *Interconnection: An Economic Impact Analysis*, Table VII–7, "Connecting Units Furnished by Bell System (FCC Data)," p. 134. (Summarizes data from *Telecommunications Reports.*)

6/73–12/74: *Telecommunications Reports*, "Total of Bell Connecting Units," various issues.

6/74–6/75: AT&T.

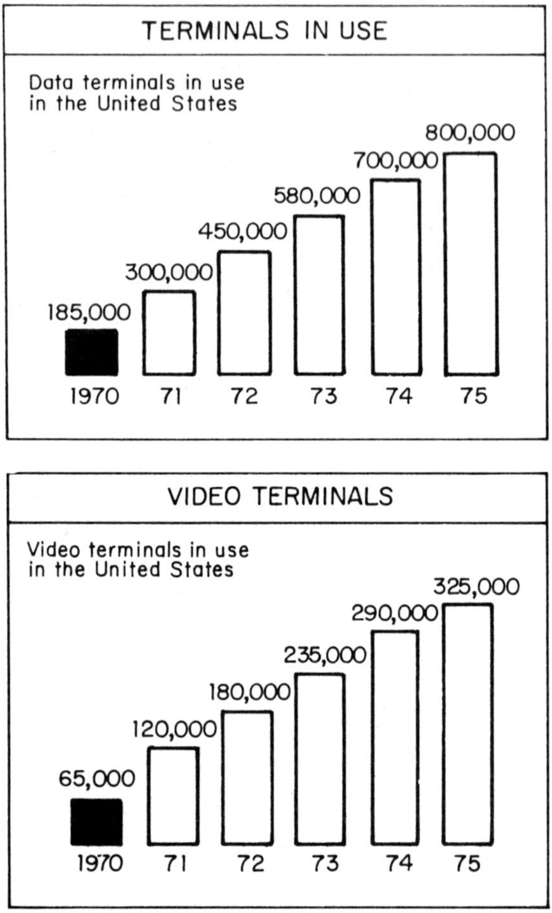

(a)

(b)

Figure 1–35. Projected Demand for Data Transmission Services

Dramatic projections of increases in the use of (a) data terminals (e.g., teletypes) and of (b) video terminals (cathode ray tube displays), (c) estimated Bell System annualized revenues from data communication service from 1962 to 1973, and (d) projected future growth all show rates 5 to 10 times the growth rate of the GNP.

Sources: (a), (b): U.S. Congress, Senate, Committee on the Judiciary, Subcommittee on Antitrust and Monopoly, Hearings on S. 1167, Part 2, "The Communications Industry," July 30 and 31; August 1 and 2, 1973, p. 1224.

(c): American Telephone and Telegraph Company, personal communication.

(d): U.S. Congress, Senate, Hearings on S. 1167, p. 1224. (Original source: Frost and Sullivan).

Year	Estimated Annual Revenues ($ Millions)	Annual Growth (%)
1962	175	-
3	190	8.6
4	205	7.9
5	230	12.2
6	260	13.0
7	300	15.4
8	355	18.3
9	440	23.9
1970	535	21.6
1	635	18.7
2	750	18.1
3	935	24.7

(c)

(d)

Year	Number of Terminals	Averaged Revenue per Terminal	Annual Data Revenues ($ Billions)	Percent Growth for Data Revenues
1970	280,000	$1,800	0.50	50%
1	384,000	1,800	0.69	38%
2	545,000	1,750	0.96	38%
3	720,000	1,700	1.23	29%
4	960,000	1,650	1.58	28%
5	1,250,000	1,600	2.00	27%
6	1,620,000	1,550	2.50	25%
7	2,120,000	1,500	3.20	27%
8	2,650,000	1,450	3.86	21%
9	3,300,000	1,400	4.60	20%
1980	4,100,000	1,350	5.55	20%

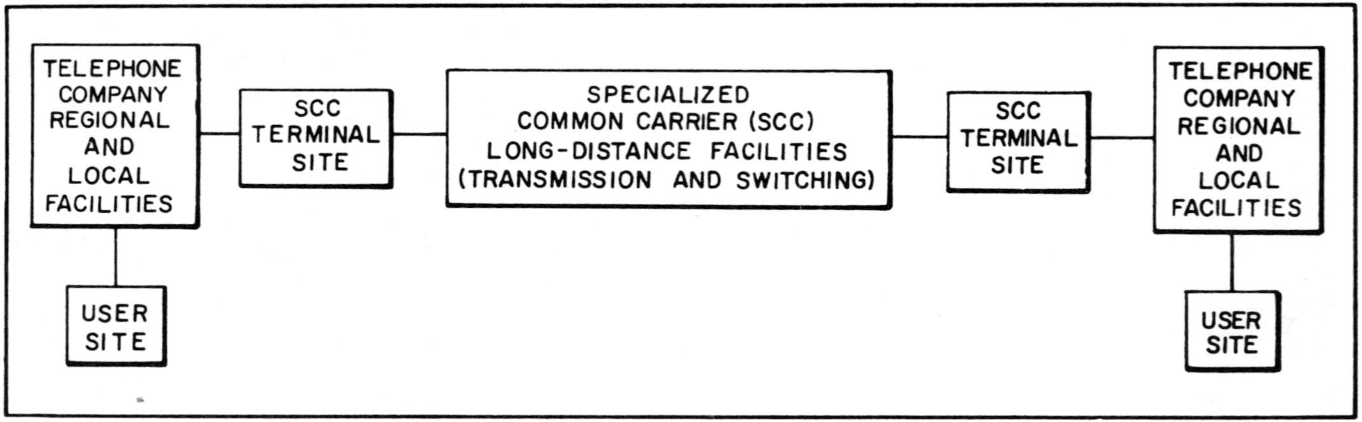

Figure 1–36. Competing Long Distance Networks

Simple specialized common carrier systems link terminal sites typically located in major cities through transmission facilities they have built themselves or, in some cases, leased from regular terrestrial or satellite carriers. Regional or local distribution from these sites is through the facilities of the regular carriers. Datran's facilities include its own switching system, located in Chicago. Satellite Business Systems' network proposed in December 1975 envisages less reliance on regular carriers by providing customers with receiving earth stations on their own buildings.

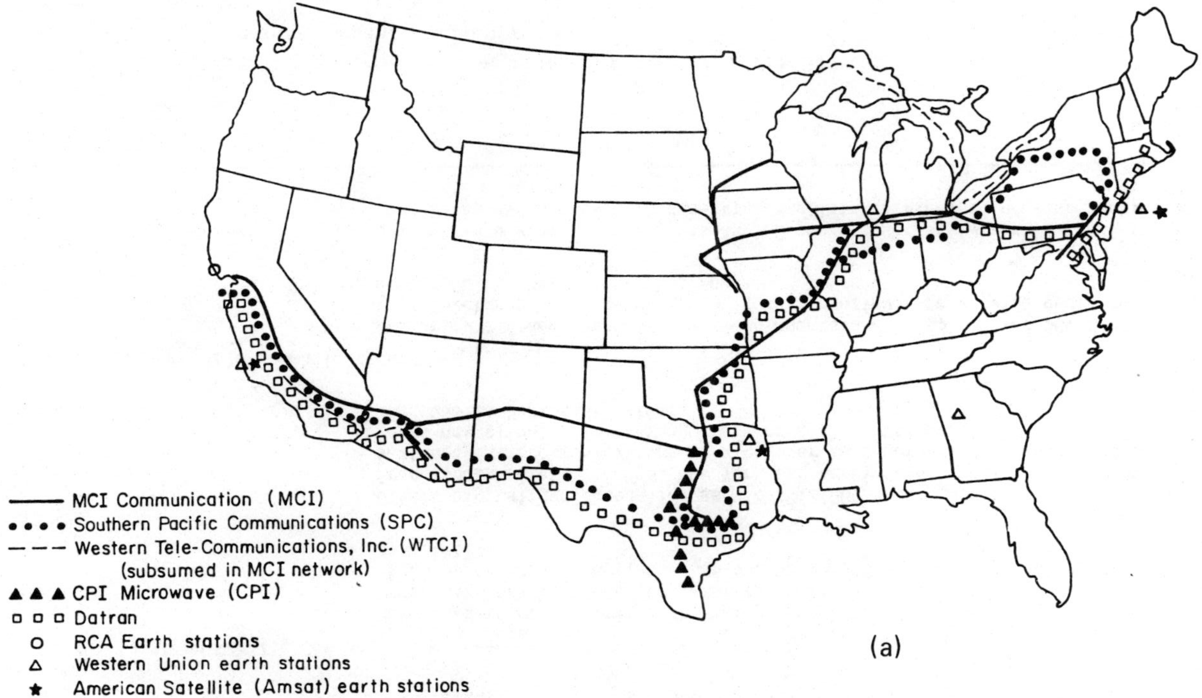

MCI: Coast-to-coast, uses WTCI (Ariz/Calif) and WUT (LA–SF). SPC: Own microwave used coast-to-coast, augmented by WUT and Westar circuits. DATRAN: Uses DDS to Northeast cities, SPC to West.

Figure 1–37. Specialized Common Carriers and Their Routes

- MCI Communications -- first to enter service in 1972; now serves 34 cities on routes including Houston, St. Louis, New York, Washington, D.C., and between Houston, Los Angeles and San Francisco. Offers range of analog dedicated channels from less than one voice grade channel to 240 VG at rates generally lower than AT&T's; 2100 VG channel capacity on network, with New York to Chicago link recently expanded to 5600; 1200 from WTCI, 60 from WUT between LA-SF.

- Southern Pacific Communications -- Has coast-to-coast service initially assembled through purchase or lease of other carriers or other carriers' facilities; offers range of dedicated channels (analog) from less than one VG to 240 VG at rates generally less than AT&T's.

- DATRAN -- Offers digital dedicated line services (2400, 4800, 9600 bps) between San Francisco, Dallas, Chicago, New York and Washington. Also offers circuit-switched data communications service over this network; recently reduced rates now slightly lower than those of AT&T's Dataphone Digital Service. Uses DDS to Northeast cities.

- Western Tele-Communications -- Primarily a video carrier in western mountain states; leases SCC facilities to MCI between Los Angeles and Arizona; continues to provide dedicated network to U. S. Navy along this route.

- CPI Microwave -- Originally a video carrier; as an SCC, serves cities in Texas with interconnection agreements with domestic satellite carriers to provide service to points outside Texas.

(b)

Fig. 1–37. cont.

(a) Specialized common carriers' routes, as described by tariffs in mid 1975.
(b) Summary data on specialized common carriers.

Source: (a) and (b) Arthur D. Little, Inc.

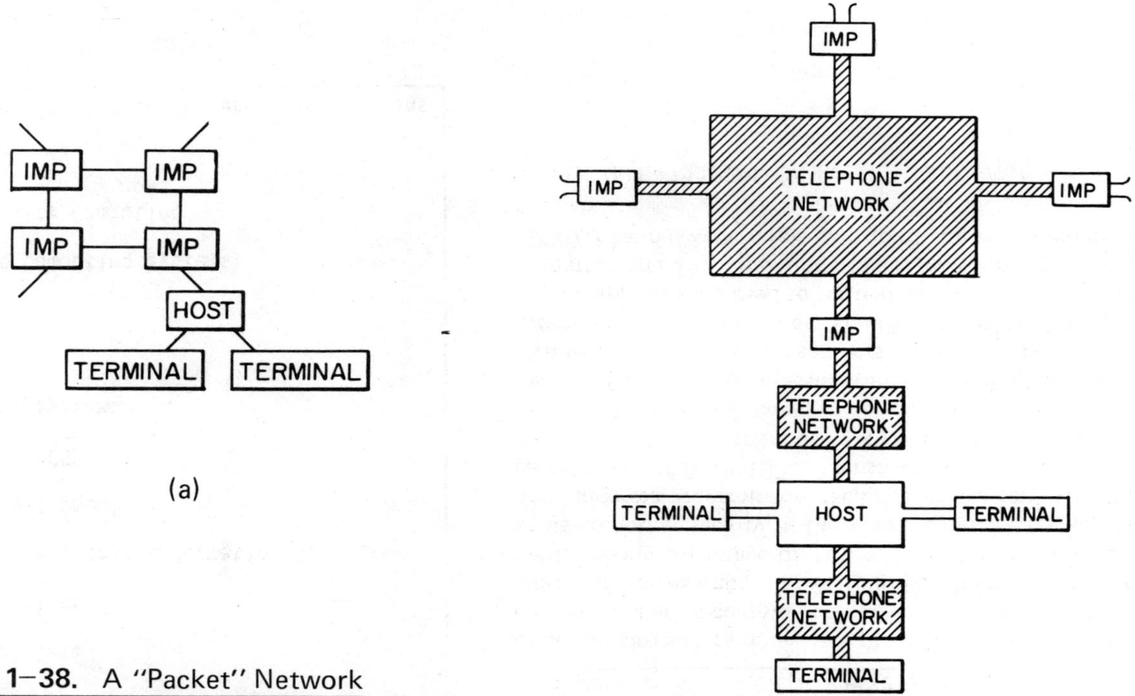

Figure 1–38. A "Packet" Network

In a packet network (a) each host computer (HOST) serving several terminals is connected to the transmission network through an interface message processor (IMP). An IMP formats messages into packets, each containing routing information. Each packet is routed from IMP to IMP toward its final destination. Alternate routings are made available to lower the probability of outage and to optimize traffic loading around the network. In actuality, connections among IMPS, HOSTS and terminals are made through telephone network facilities (b). Are the HOSTS outside or inside the telephone network?

* Computer Sciences
* ComShare
* Control Data
* Cyphernetics
* General Electric Information Services
McDonnell Douglas
National CSS
On-Line Systems
Rapidata
* Tymshare
United Computing Systems
University Computing

* International networks and/or operations

(a)

Figure 1–39. Teleprocessing Services

Some teleprocessors retail data processing services to multiple customers sharing common computing facilities (a). G.E.'s network (b), for example, serves 5,000 customers on four continents, taking advantage of 18 time zones to smooth workload at its computer facility in Cleveland, Ohio. Terrestial, satellite and undersea transmission facilities are leased from regular carriers. As of April 1973, Tymshare's Tymnet connected 54 cities with 37 computers using 40,000 miles of leased private lines. The boundary between telecommunications (regulated by the FCC) and teleprocessing (unregulated) is hotly disputed. Tymnet has joint use agreements with other companies that access their own computers via Tymnet; in this case, "Tymnet's computers are used to route data but do no processing"—a matter the FCC has undertaken to scrutinize in Docket 20097 dealing with resale and joint use of common carrier services.

47 FCC 2d 647 (1974), Docket No. 20097, *Notice of Inquiry.*

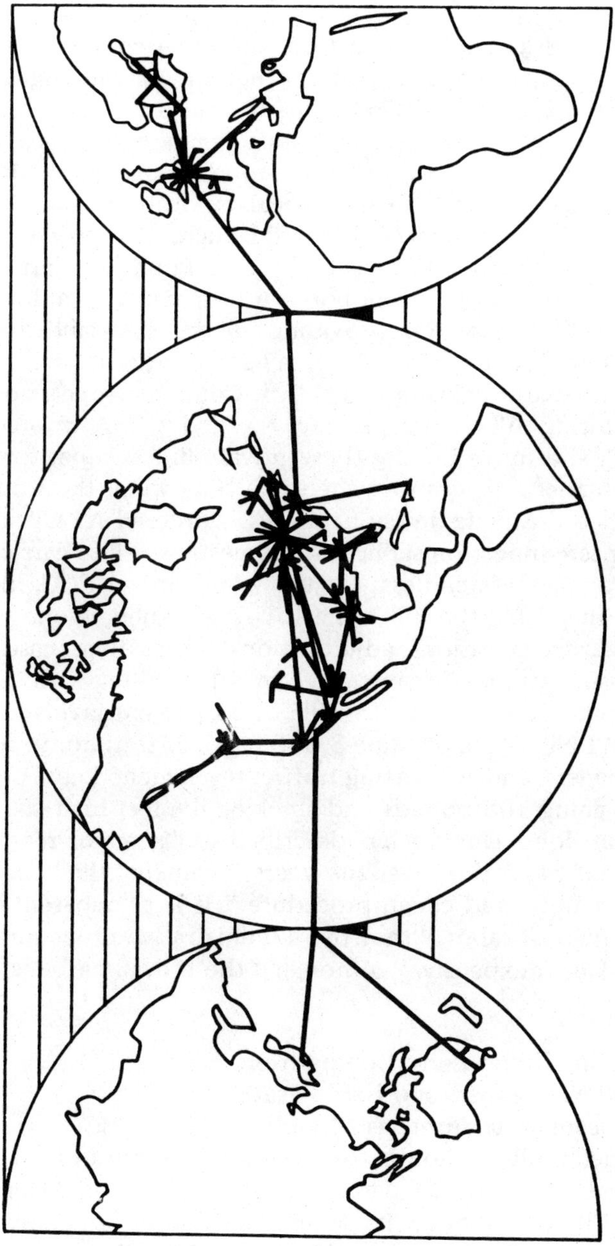

(b)

7. HI-LO AND INTERCONNECTION: THE NEWS MEDIA UNDER CROSSFIRE

The war over SCCs, interconnection, and other competitive pressures is fought on many fronts, with the news media among the participants. Tactics include political maneuvers in federal and state legislatures, legal moves in federal and state regulatory bodies, as well as full-blown litigation in the courts, economic moves in the marketplace, and searching for technological innovations (see Fig. 1–23).

For example, in July 1974, Richard Hough, the president of AT&T Long Lines, testified before U.S. Senator Philip A. Hart's Subcommittee on Antitrust and Monopoly on how many local loop connections the Bell System was providing for the SCCs and their private line customers. At the same time, AT&T was awaiting the outcome of its unsuccessful challenge in a U.S. Court of Appeals on an FCC order requiring AT&T to provide FX and CCSA interconnection to the SCCs, a move linking these private line competitors into the switched services. And about the same time when the Court of Appeals handed down its decision, the FCC ordered AT&T also to provide full interconnection services to domestic satellite carriers.

The examples are endless, for the world of the FCC, like that of the CAB, the FDA, the FTC, the ICC, and others is one of endless, massive hearings, petitions, adjudications, proceedings, cases, appeals, and reconsiderations. For instance, an AT&T filing with the FCC on increasing its rate of return to 10½ or 11 percent involved a submission, by AT&T alone, of some 8,500 pages of testimony, statements, working papers, and supporting traffic, revenue, and cost data—documents weighing 108 pounds and stacking 4½ feet high [52]. Reality belies what John Dunlop has described as "a major reason for the attraction of regulation over the years," namely "the belief that it is a speedy, simple, and cheap procedure." It is as apparent here, as in Dunlop's field of labor, "that the administrative procedure is by no means fast or inexpensive" although "the prevailing belief is that it is" [53].

The Pricing Response to Competition: Death to Small Newspapers and Radio Stations?

On the economic front, as of early 1974, the SCCs were quoting 75¢ to 95¢ a mile a month for the first 500 miles of a voice grade private line, or 52 to 66 percent of AT&T's $1.44. The FCC had required AT&T to request its permission before filing any interstate tariffs, instead of, as usual, reviewing filed tariffs and exercising its discretion to suspend them for 90 days at most. This requirement was challenged by AT&T and overruled by the U.S. Court of Appeals [54]. Meanwhile, AT&T had initiated a price response that illustrates the profound effects that competition can have on pricing as practiced by a regulated monopoly. Hitherto (as outlined in Section 5) all pricing had been related to costs principally through the overall "costs = revenue requirements" equations.

Absent competition, the traffic can bear value-of-service pricing for basic business telephone service and state regulators can urge or approve it in exchange for the lower residential rates (see Fig. 1–24). Absent competition, interstate private line services can be priced in keeping with a cost averaging viewpoint and the policy defended by regulator and regulated as a fair way to keep prices independent of demographic accident: prices vary mainly with crow's flight distance between connected points, thus averaging out any differences that differing costs of serving different equidistant points might induce in cost-based pricing. A competitor would seize on the opportunity to capture business by pricing near his cost on routes where average monopoly prices are set much higher than unaveraged costs to compensate for the routes where averaged prices fail to cover unaveraged costs. As Figure 1–8 showed, a technical argument can be made that economies of scale yield unit costs that are much lower where traffic is high than where traffic is low.

The SCCs, in a move denounced by AT&T as "cream-skimming" settled on high density private line routes (see Figs. 1–25b and 1–37), and undercut AT&T's prices. AT&T responded by departing from price averaging toward a cost-based tariff, with lower prices for high traffic density routes and higher prices for low traffic density routes. Under this so-called Hi-Lo tariff as originally proposed, $1.44 per mile per month would decrease to $.85 on the favored routes and increase to $2.50 on the others [55]. After many rounds in the FCC and the courts, the Hi-Lo tariff went into effect in June 1974, but continued to be challenged in a storm of mutual charges of cream-skimming, cost distortion, and unfair competition. A competitor's view of this tariff's effects is shown in Figure 1–40.

The actual impact of the Hi-Lo tariff on news media is far from clear. Newspaper publishers have challenged the tariff vigorously (Fig. 1–41), claiming unfair discrimination against them, and suggesting that there would be dramatic effects on who subscribes to news wires, hence on the news that everyone reads or watches. Wire traffic may be concentrated, but all the public is not. The news media use a lot of those low traffic routes. But are the news media merely crying wolf? UPI, for example, found a way to reduce to $5,500 per month the claimed $108,000 per month cost increases entailed by the Hi-Lo tariff, without changes in the effective capabilities it purchased.

The way out, for UPI and others, was the Telpak exception. The Telpak private line tariff offers private line service in bulk to those with high capacity requirements; under this tariff, prices per line are lower than if the lines were leased one by one under the "retail" private line tariffs. With elegant euphemism, the Telpak exception permits so-called end links to be tied to a short segment of Telpak lines, without any restriction on the end link mileage; and, the end links are priced at pre Hi-Lo rates. The combination thus ingeniously bypassed the high rates that would otherwise apply to UPI's many lengthy "end links" along low density routes.

However, in September 1975, the FCC declared the Telpak exception to be unlawfully discriminatory [56]. It found no hard evidence for claims that many small newspapers and radio stations would reduce or eliminate their wire service subscriptions because of Hi-Lo rates. Indeed, the only survey made available to the FCC "was submitted by the National Association of Broadcasters and consists of a poll of a sample of radio stations. The stations were told the purpose of the survey, and the NAB General Counsel admitted at the oral argument this fact may have biased the results" [57].

The FCC ordered AT&T to file tariffs eliminating the discriminatory Telpak exception. However, it also undertook to "initiate an investigation into the need for preferential press rates for private line services" [58]. Claiming "serious and irreparable injury" with no one "standing to gain any meaningful benefit" [59], the news interests called on the FCC and the courts to provide interim relief pending the ultimate settlement of the Hi-Lo matter. AT&T filed revised tariffs to be effective in November 1975, but their effectiveness was stayed by an order that the U.S. Court of Appeals, District of Columbia Circuit, issued on November 18, 1975 and that remained in effect as of April 1976. And so one typical round spins off others.

In January 1976, the FCC decided to declare the entire Hi-Lo tariff unlawful, null and void, an action not nearly as drastic as the legal terms sound. Declaring that "the Hi-Lo concept is, as a general principle, a valid type of competitive response by Bell" [60], the Commission just "found the record insufficient to support AT&T's claim that the specific rates and conditions in the Hi-Lo tariff were a reasonable application of [the] principle" [61] of departing from nationwide rate averaging.

In April 1976 AT&T filed a new tariff, baptized MPL for Multischedule Private Line service. In effect, the new tariff included rate steps that sharply reduced longer haul prices per mile and sharply increased shorter haul prices per mile over what they were under the Hi-Lo tariff's flat per-mile rates. The Hi-Lo concept was retained in defining distinct rate schedules, with lowest rates between places linked through high capacity circuits (category A), highest rates between all other places (category B), and intermediate rates between places in Category A and places in Category B [62].

The Hi-Lo and MPL tariffs are not the only examples of moves away from averaging justifications for private line prices and toward cost-based pricing. On yet another merry-go-round, tariff changes made in 1973 [63] included temporary (to 12/75) increases in the prices for "one-shot" or occasional communications channels for television program transmission over the prices for long term lease of those channels. This change might further constrain the news producer's decisions as to timeliness versus cost in choosing to have stories sent to news headquarters either by wire as a stream of electronic signals or else by plane as a roll of film or videotape depending on how far the remote location is from his network's "spine." Then again, it might not. Whatever their specific effects on news media or on any other user category, the moves toward cost-based pricing stimulated by the emergence of competition in a hitherto monopolistic arena, inevitably called to center stage all the thorny questions of cost definition, sketched in Section 5, that hitherto lay hidden in the wings [64].

The following illustrates the tenor of the arguments over cost definition, including appropriate accounting periods. Cross-subsidization of one service by another is claimed if some costs of the first are borne by the second. Predatory pricing is claimed if price is below cost. One side in an argument over cross-subsidization or predatory pricing therefore has a stake in demonstrating that "true" costs are lower and the other that they are higher than what the opponent claims. In defending private line tariffs, AT&T prefers considering what economists call "incremental costs" over using "fully distributed costs" as advocated by its competitors (Fig. 1–42). As illustrated by service B in Figure 1–42, incremental costing would justify prices yielding as low as $5 million of revenues, while the fully distributed costing method would put the floor at $6.7 million. A competitor who can produce a service like B at an incremental cost of $5.8 million and at a fully distributed cost of $6.1 million could thus undercut AT&T's prices if fully distributed cost is the floor but not if the floor is incremental cost. Competition on the terms of this example is clearly inefficient from a purely economic viewpoint; the inefficient competitor produces at a cost of at least $5.8 million some or all of what could be produced at an expenditure of only $5 million of society's resources.

The economic theories that regulators and other parties might invoke in particular cases depend less on what might be shown by simple, static and clear cut examples than on the complex, shifting, historically cloudy and prospectively indeterminate facts of actual cases and on the parties' policy objectives of the moment. One case in point is the January 1976 recommendation by the Chief of the Common Carrier Bureau to the FCC that fully distributed cost (FDC) principles be followed in pricing Telpak and other services in the competitive private line arena [65]. Bowing to the judicially sanctioned "relative use" principle, but otherwise understating the arbitrariness of it all, the recommendation describes common costs as "allocated among the services according to statistical and/or judgmental determinations as to the proportionate use which each service made of the facilities, personnel and operations supported by such costs" [66]. In the very next sentence, certainty erupts in a statement that "at the conclusion of this process, it is possible not only to compute a rate of return for each service, but also to establish *definitive* unit costs for each of the services . . . " (emphasis added).

Later, it is conceded that "the basic issue regarding the allocation of common and joint costs appears to be the inherent arbitrariness of such decisions" but it is argued that "the LRIC [long run incremental cost] approach may be said to be more arbitrary than that of FDC" [67]. The "less artibrary" FDC method recommended to the Commission is said to follow "the actual or relative use concept of the jurisdictional separations manual," namely the Alice-in-Wonderland formulas of the Ozark Plan described in Section 5. The justification offered is that "within the context of the adaptation of these procedures to special services, *it seems evident to us* that the principles contained in the Manual are suitable for determining costs and revenues assigned to various service categories" (emphasis added) [68].

By contrast, in November 1975 the chairman of New York State's Public Service Commission, while also praising competition as the "enemy of value of service pricing," saw competition as an enemy that "forces all prices to incremental cost" [69]. His concern of the moment was a major rate case involving the New York Telephone Company's terminal equipment services; his staff had determined that historical fully distributed costs for these services were lower than incremental costs at that moment. He noted that "the interest of the telephone companies . . . is to hold on to [the terminal] business by holding prices *down*, below incremental cost if necessary," a tendency to which "the present method of separations provides a powerful reinforcement" [70].

Such ironic contrasts are invisible within any one of the multitude of fragmented regulatory proceedings. Nor is consistency a virtue when trying to win cases. Strong blasts of authority and certainty propel the economic arguments and rebuttals shot back and forth by the clattering squads of accountants, economists, engineers, lawyers, scientists, and other experts manning the trenches of the various parties. Ineluctably, however, all this expertise bears down on a game where the size of the major "costs = revenue requirements" pots must be set by political decisions comprising not only the jurisdictional separations processes described in Section 5, but also policy as to rates of depreciation, fair rates of return on debt or equity capital and numerous other factors. As Figure 1–43 shows, there are conditions in which the sharing of common costs cannot reach a purely economic equilibrium even in theory; the intervention of political processes is essential.

What conflicting expertise contributes to the fragmented adjudicatory and rule making processes of administrative agencies and the courts (Fig. 1–44)—and what it is sought for—is not truth as revealed to experts, but rather political and judicial maneuvering room spacious enough to accommodate the demand of both potent constituencies or clienteles and of judicial review.

Technology and Competition

Alongside the foregoing skirmishes, AT&T's research arm, Bell Laboratories, and Western Electric have not been idle on the technological front. A standard Bell transmission mastergroup, encompassing 600 voice-grade channels, uses a bandwidth of 2.52 Mhz between .564 and 3.084 Mhz [71], leaving empty the spectrum below .564 Mhz. With a newly developed modulation system, labeled DUV for "Data Under Voice," the Bell System can fill that empty band in every radio channel in the long haul transmission system with a digital signal stream of 1.5 Megabaud, equivalent to a digital T1 channel (Fig. 1–33). In early 1976, this DUV technology was the principal basis for a new digital transmission private line service that AT&T calls DDS, for Dataphone Digital Service.

Figure 1–40 shows one view of the competitive effect of DDS. As might be expected, arguments over DDS are rife in all the usual places. In December 1974, over two years after AT&T first applied, the FCC approved its offering DDS among Boston, Chicago, New York, Philadelphia, and Washington, D.C. at special low DDS rates. Service to 19 other cities was authorized at the same time, but with rates pegged at the high low density rates of the Hi-Lo tariff. The authorization was appealed by Datran before the U.S. Court of Appeals for the District of Columbia and the FCC continued to examine DDS rate principles under Docket No. 20288. By April 1976, Datran, increasingly dependent on AT&T for facilities to extend its own services, had withdrawn its appeal and DDS rates were in effect for all 24 cities. As might also be expected, there are disputes over whether DUV costs should be counted as incremental or as fully distributed; over whether, in either case, prices cover costs; and over whether pegging DDS rates high in some cities unfairly protects some of the competitors.

The Hi-Lo and MPL tariffs might hurt news media; but then again the Telpak exceptions, court-ordered stays or special press rates might keep riding to the rescue. News media might gain in one pocket by taking advantage of SCC rates or of responsive lower AT&T tariffs; but then again they might lose in another pocket if, within fixed overall "costs = revenue requirements" equations, lower private line prices mean higher MTS or other prices.

Similarly ambiguous primary effects and still less determinate secondary ripples arise as a consequence of competition over the "interconnect" market of terminal and like devices. As newspapers and wire services rely more and more on compunications technology, surely they will benefit if competition drives down the price of these devices. But, it is said, major ripples will offset this benefit by raising the prices of MTS and other services.

John Kenneth Galbraith wrote that "the study of money, above all other fields in economics, is the one in which complexity is used to disguise truth or to evade truth, not to reveal it" [72]. The technical complexity of compunications systems is real enough, as is the gap between economic theory and the realities of practice. But there is no truth to disguise. There are only the timeless questions of who gets what, when, how from the complex games we have described and by whose say so. The concluding sections summarize our findings by framing these universal questions as they apply to these specific games.

(Text continues on page 122 . . .)

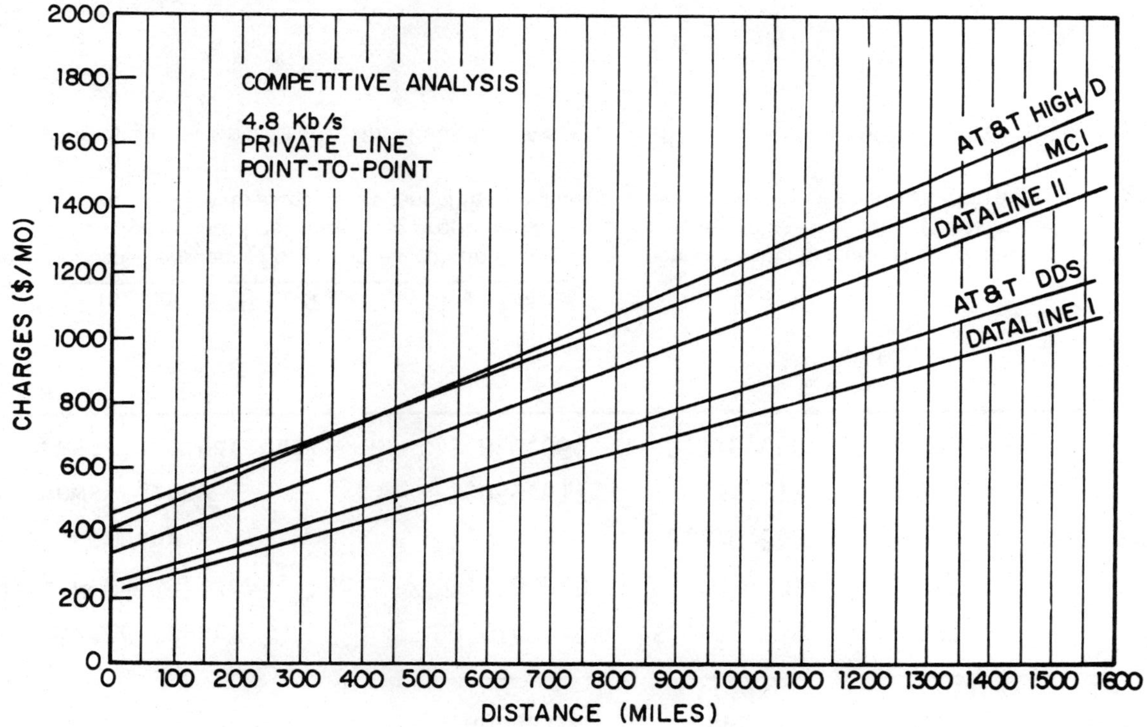

Figure 1–40. Price Competition in Private Line Services

The interleaving of AT&T's high density (low price) and DDS tariffs, with those of MCI and Datran (Dataline I and II) illustrate a competitor's view of the competitive interplay with AT&T.

Source: Data Transmission Company.

	Present Annual Cost	HiLo Annual Cost	Annual Cost Increase	Percent Increase
The Associated Press	$3,414,511	$4,485,713	$1,071,202	31
Commodity News Services, Inc.	1,493,808	2,263,537	769,728	52
Dow Jones Company	1,391,995	2,526,559	1,134,564	82
Reuters News Agency	1,059,617	1,634,317	574,700	54
United Press International	3,820,056	5,116,056	1,296,000	34

Figure 1–41. Impact of Hi-Lo Tariff on News Media

In 1974, the American Newspaper Publishers Association anticipated increases ranging from 31 to 82 percent in the prices paid by news wire services for private lines under the proposed Hi-Lo tariff. As heavy users of private lines in otherwise low density areas and of many short links with associated channel terminal costs, they complained of bearing a disproportionate share of the high price side of the Hi-Lo tariff.

Source: Submission by the American Newspaper Publishers Association, FCC Docket No. 19919, July 18, 1974, p. 14.

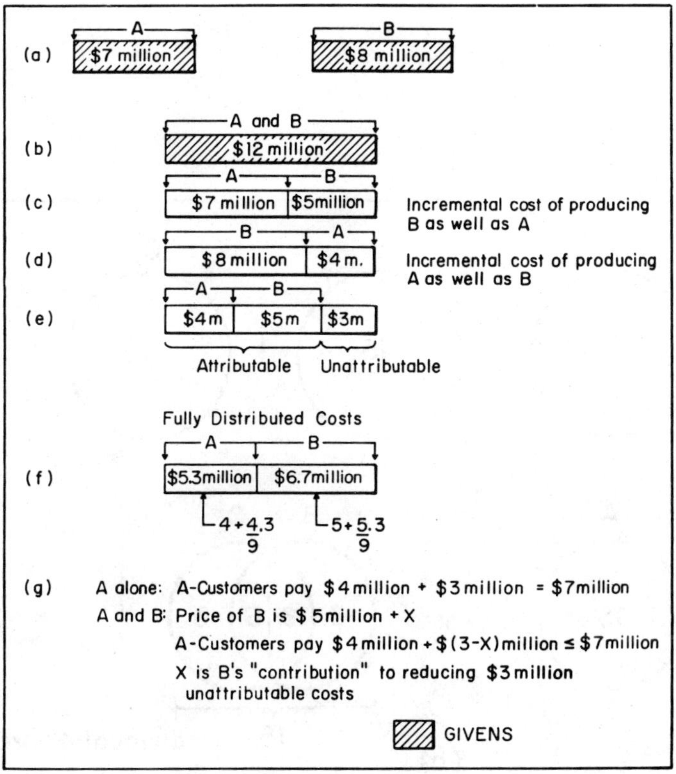

Figure 1–42. Fully Distributed and Incremental Costs

Assumed costs of producing A and B separately and jointly are shown in (a) and (b). The incremental costs of joint production are shown in (c) and (d). Assuming that the sum of the incremental costs is a fair measure of attributable costs, remaining unattributable costs are $3 million (e). This assumption is tenuous in practice; if there is no plant at all, all costs are engineering estimates and the decision as to which product is the add-on is arbitrary; if the plant for A exists, the incremental cost of B lends itself to engineering estimates, but the incremental cost of A (d) given a plant for B is entirely hypothetical. An arbitrary decision to allocate unattributable common costs by the ratio of attributable costs is reflected in (f). Shown in (g) are arguments by proponents of incremental costing to the effect that customers of the older service A are necessarily better off so long as customers from the newer service B pay at least the incremental costs of B. An estimate of unattributable costs is not essential if this argument is accepted, but it is required if comparisons with fully distributed costs are to be made. Where, as is often the case in telecommunications, the new service shows cross-elasticities with other services, the definition of incremental cost and the measurement of any burden or cross-subsidies are further complicated.

Source: W.J. Baumol and A.G. Walton, "Full Costing, Competition and Regulatory Practice," 82 *Yale Law Journal* 639, 1973.

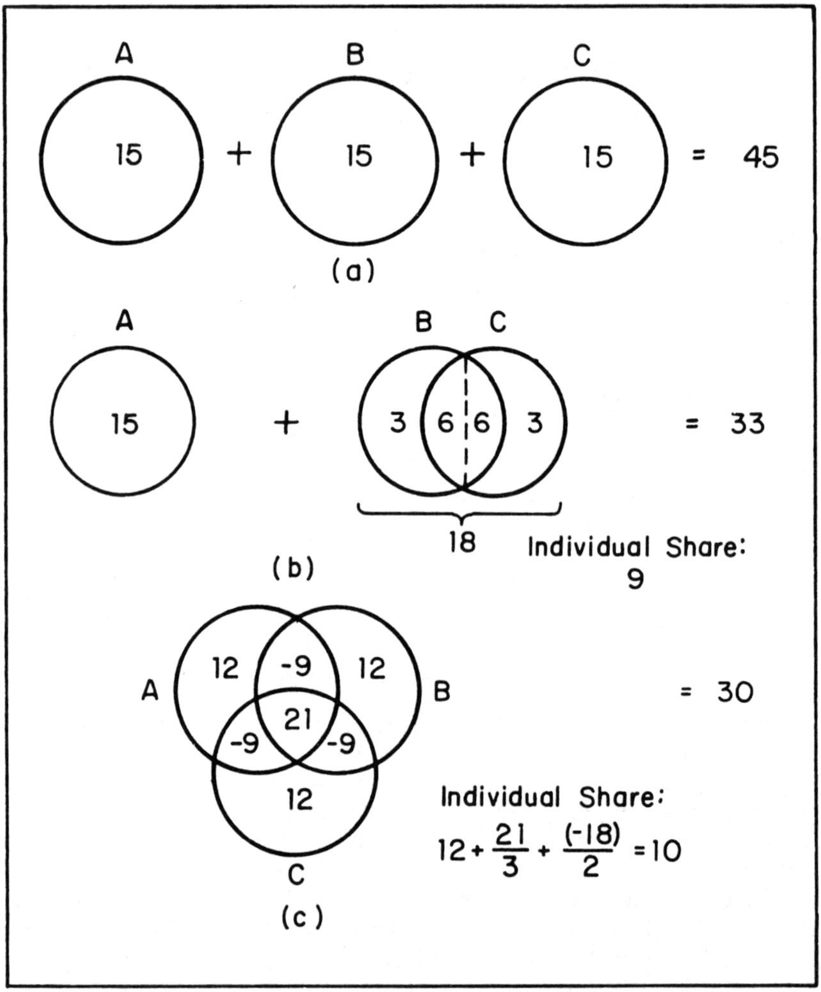
A
B
C
15
+
15
+
15
= 45
(a)
A
B C
15
+
3 6 6 3
= 33
18
Individual Share:
9
(b)
12
-9
12
A
B
= 30
21
-9
-9
12
C
Individual Share:
12 + 21/3 + (-18)/2 = 10
(c)

Figure 1–43. Theoretical Instability of Common Cost Sharing

The process of sharing common costs can be fundamentally unstable even if the overall costs of service decrease from 45, as in (a), to 33, if two services share common plant as in (b), and to 30 if all three share as in (c).

In (b), A has an incentive to woo either B or C away from the other, but this incentive then passes to the remaining singleton. Moving from (b) to (c) means that any pair of services must contribute no more than 18 to feel better off. But the singleton must then contribute at least 12 for the total to be 30. A fair solution would then require every participant to contribute at least 12, hence any pair at least 24, contrary to the assumption. Mathematically, the absence of incentive for *every* participant to move from (b) to (c) is reflected in the absurd negative joint cost elements in (c); or, in terms of game theory, in the absence of a "core."

Looked at another way, the diagrams show how dependent the "incremental" costs of any service provided jointly with other services are on the order in which the costs are determined. The incremental cost of adding any one of the services in (a) to any other in order to achieve the savings in (b) is 3. However, one and only one of these services can claim that incremental cost; which one does so is indeterminate without intervention from outside the theory.

External intervention would be required to assign equal shares of 10 to each participant in order to attain the overall lowest cost of 30.

Source: G.R. Faulhaber and E.E. Zajac, "Some Thoughts on Cross-Subsidization in Regulated Industries," Bell Laboratories Economics Discussion Paper, No. 48.

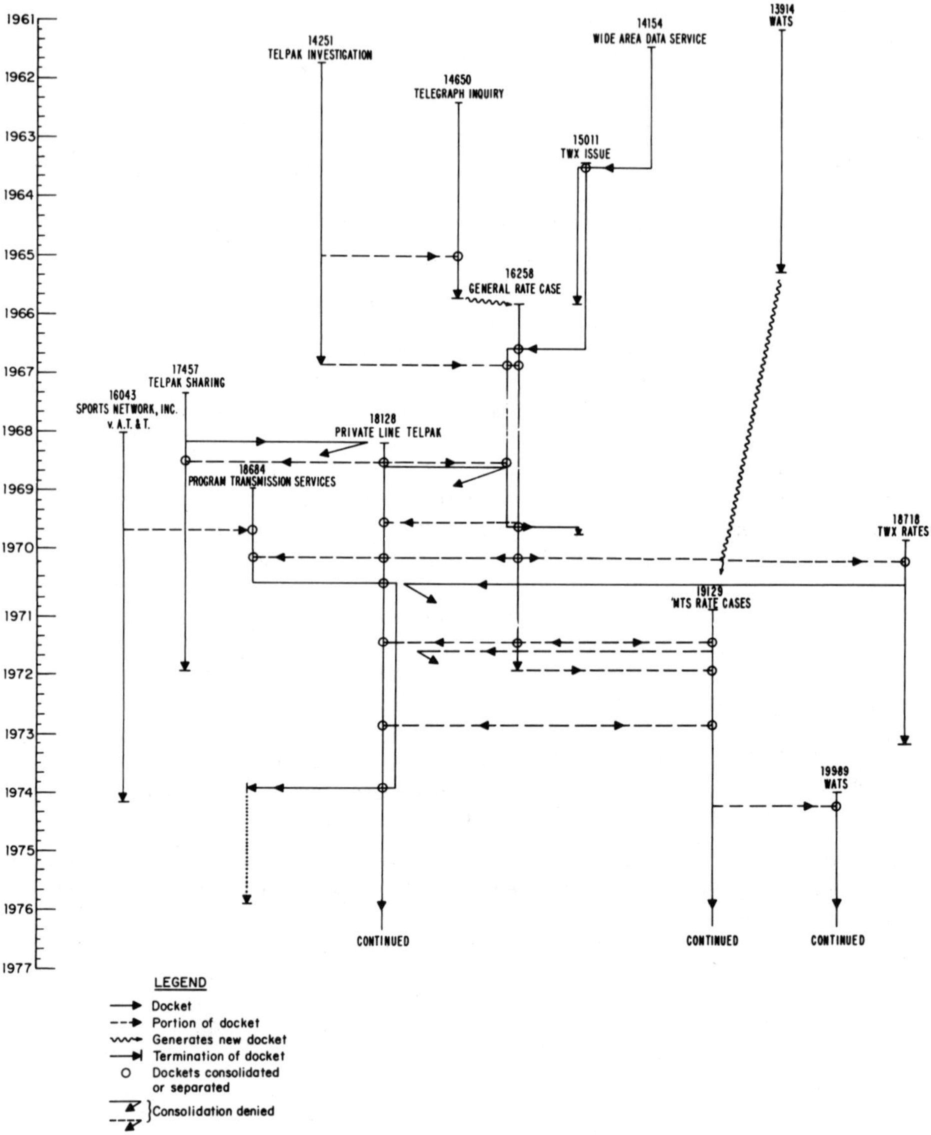
1961
1962
1963
1964
1965
1966
1967
1968
1969
1970
1971
1972
1973
1974
1975
1976
1977
13914
WATS
14154
WIDE AREA DATA SERVICE
14251
TELPAK INVESTIGATION
14650
TELEGRAPH INQUIRY
15011
TWX ISSUE
16258
GENERAL RATE CASE
17457
TELPAK SHARING
16043
SPORTS NETWORK, INC.
v. A.T. & T.
18128
PRIVATE LINE TELPAK
18684
PROGRAM TRANSMISSION SERVICES
18718
TWX RATES
19129
MTS RATE CASES
19989
WATS
CONTINUED
CONTINUED
CONTINUED
LEGEND
Docket
Portion of docket
Generates new docket
Termination of docket
Dockets consolidated
or separated
Consolidation denied

Figure 1–44. The Regulatory Web

Controversy over TELPAK tariffs dates back to their inception in the early sixties; it is interwoven with proceedings on other services and one on general rate principles (Docket No. 16258). The Common Carrier Bureau Chief's recommendation that costs of TELPAK and other private line services be determined according to fully distributed costing principles as embodied in the Ozark Plan issued in Docket No. 18128. The FCC's decision to end the TELPAK exception was set forth in the Hi-Lo proceeding (Docket No. 19919), which is not shown on the chart. Summaries of some key proceedings are in *Agenda for Oversight: Domestic Common Carrier Regulation* (U.S. Congress, House of Representatives, 94th Congress, 2nd Session, Committee on Interstate and Foreign Commerce, Subcommittee on Communications, Subcommittee Print, April 26, 1976).

8. WHO GETS WHAT

In the mid sixties, the Ford Foundation suggested that TV network use of a proposed television transmission satellite be taxed by setting prices for its use higher than the economists' most efficient prices, but below the point where it would pay for the networks to set up their own satellites. The proceeds were to be dedicated to the development of "merit" programming.

Massachusetts Fair Share, "a non-profit citizens' action organization fighting for basic economic justice for all low and middle income families" [73], questioned, in an article in its magazine, *The Fair Fighter*, whether the $600,000 the New England Telephone Company spent in charitable contributions was an expense the customers should pay for. However, in the same article, Fair Share also proposed that "every family should be able to get a fixed amount of telephone service at a rate they can afford—especially senior citizens, people with health problems, and families on a fixed income."

Such proposals, like the games described in the preceding sections, affect everyone directly and not only through the news media. The players in these games are not invisible; but they do not necessarily see each other, nor are they necessarily charged to look. For example, having concluded that interconnection is responsible for price increases affecting "the vast majority of New York Telephone Company's subscribers," a 1975 staff study for the New York Public Service Commission acknowledged, however, "the possibility that, as a result of interconnection, some economic benefit may accrue to telephone subscribers from sources outside the focus of this study. It was not our purpose to analyze or measure that possibility . . ." [74].

Even on the federal level, where the perspective is supposed to be broader, it was only in 1974 that the FCC found it expedient to open an inquiry on "Economic Implications and Interrelationships Arising from Policies and Practices Relating to Customer Interconnection, Jurisdictional Separations and Rate-Structures" [75]. The regulatory purpose of the FCC, as specified by Congress in the Communications Act of 1934, is "to make available, so far as possible, to all the people of the United States a rapid, efficient, nationwide, and worldwide wire and radio communication service with adequate facilities at reasonable charges" [76]. This mandate sounds narrowly technological and economic but, after forty years, it is clear that the FCC and the other players listed in Figure 1–23 in fact make broad social policy. This is not unusual, since the courts generally give an administrative agency "great deference in the construction of its own statute" [77].

Also, "the fast moving nature of communications technology, and the loose language of the statute, provide some basis for believing that Congress intended to allow the FCC a wider latitude than it typically granted to other agencies" [78].

Formal Congressional interventions have been rare since the 1930s, mainly affecting the extension of telephone service to rural areas. As of 1949, the limits of "so far as possible" were 60 percent of households nationwide (see Fig. 1–5) and 42 percent of farms [79]. By 1971, a rising tide of national affluence and the 2 percent loans that the Rural Electrification Administration had made since 1949 had helped raise farm penetration to 83 percent, compared to over 90 percent nationwide. Congress then supplemented REA loans through a new Rural Telephone Bank eventually to be owned and controlled by its patron telephone companies in what the minority termed a "multi-billion-dollar banking boondoggle" [80].

The watchword was not only extension but also upgrading of service since "busy rural people, no less than city people, count single-party telephone service an essential rather than a luxury" [81], an observation akin to those made in favor of Rural Free Delivery, when that was a controversial postal policy question late in the nineteenth century. The majority noted that single-party lines "are also basic to new communications services which are ready for wider application" [82]. The point that the main issue was who gets what, was made by the minority, which opposed what it called a boondoggle on the ground that there were "better ways to meet the legitimate needs for improving telephone service in rural areas" [83] than a proposition to help rural telephone companies that "have enjoyed inordinate and, in some instances, downright exorbitant returns on capitalization" [84]. It cited 30 companies with most reasonable returns on total capital (debt and equity) ranging from 3 to 13 percent, but with leveraged returns on net worth ranging from 30 to 400 percent [85]. "Incidentally," the minority continues, "nearly three-fourths of the telephone companies which would be eligible for these windfall loans are not cooperatives. They are commercial corporations organized for profit. They seem to be doing very well, indeed" [86].

Mostly, however, telephone regulation employs the opaque legal, economic, and technical prose of administrative and judicial processes, not of the pungent idiom of the open political forum. The politics are nonetheless as real, even if more obscure.

What do some of the principal players think about who gets what? In opening the door to specialized common carriers, the FCC reopened the question of monopoly vs. competition, essentially closed for over half a century since AT&T, under Theodore Vail's leadership, sought out regulated monopoly status as a stabilizing alternative to competition. SCC proponents saw the move as aiding in the development of the new and different teleprocessing and other services they alleged to have been held back by the regular carriers' failure to provide cheap enough and reliable enough private line facilities fast enough.

SCC opponents retort that these "other common carriers" merely judiciously duplicate ordinary Bell System services to "skim the cream" off the system's most lucrative routes, while existing capacity is underused owing to the failure of glowing demand projections to materialize [87]. They see the SCCs as establishing a shadow interstate and interexchange telephone system. They foresee that the SCC's cream-skimming will undermine the statutory mandate of service to all by destroying the proceeds that averaging and value-of-service pricing for some yield toward paying for service to others. And, as is at issue in the case of MCI's Execunet Service and similar proposals by IT&T, they fear invasion of the Message Telecommunications Service, in part through what they see as abuse of their own facilities.

In the realm of interconnection of "foreign" attachments, electronics manufacturers have a stake in capturing a share of the telephone instrument and private branch exchange market (Fig. 1–45) hitherto the exclusive preserve of AT&T's Western Electric manufacturing subsidiary and of other companies which, like Stromberg-Carlson, supply instruments to independent telephone companies. They also have a stake in the rapidly growing new market for computer and other terminal equipment increasingly used for data and other services which, like UPI, depend on compunications technology. Conversely, the present communications common carriers have a direct economic stake in losing to competitors revenues from manufacturing and from renting telephone instruments and other terminal devices, including new devices that might be computer-like or communications-like according to the sweep of regulatory definition and the scope of technological innovation.

Little of substance is known about causal links among technological innovation and diffusion and their context. In April 1976, AT&T submitted to the FCC a study by Arthur D. Little, Inc. commissioned to "find and explain relationships between the innovation process and market structure in telecommunications" [88]. Finding that neither theory nor evidence support the idea that competition necessarily increases innovation, the report that "collectively, past studies provide only inconclusive and ambiguous support for any such general propositions relating total amount of innovation to market structure and regulatory practices'' [89]. Others ascribe specific importance to telephone company decisions about directions of technological innovation. For example, the chief of the FCC's Common Carrier Bureau sees the pressure on state prices as stemming not from competition, but from flat rate rather than usage sensitive pricing, wage inflation, and the "inability of the industry thus far to produce [for local service] the cost-saving technological and operational innovations characteristic of toll services'' [90].

There was, of course, much technological innovation even in a calmer 50-year interlude. The telephone company, however, could determine the nature and the pace of innovation, perhaps concentrating on long distance technologies so as to be able to keep rates low in a market that exhibited relatively elastic demand and continuing growth—in other words, where the money is. Even then, technological changes such as introducing direct distance dialing could perturb the formal separations procedures by reducing the line holding time per toll message, and with it, the proportion of plant allocated to federal rather than state jurisdictions [91].

But the introduction of competition may well have speeded up the spread of such innovations as DUV, and new Private Branch Exchange (PBX) devices relying on microprocessors and stored program facilities—microcomputers—instead of mechanical or hard-wired electronic devices [92]. And perhaps the pressure on state and local rates has prompted fresh concentration on local distribution technologies, leading to the development of such innovations as the Rural Handset—a device that encompasses solid state amplifiers within the handset to boost the volume of both incoming and outgoing telephone calls [93], thereby reducing the need for more expensive line amplifiers. Ascribing such triggering effects to competition is not inconsistent with the A.D. Little findings that "DUV could not have been developed in the early '70s without a string of underlying innovations and research results that occurred more or less uniformly over the preceding one to two decades'' [94] and that "innovations in PBX systems have been made all through the period from the 1920s to the present, with a natural decrease in activity in World War II'' [95].

These technological issues do not present themselves so much as polar choices than as matters of degree or of choice of mix, subject to perceived relationships among costs, prices and demands for services exploiting available technological capabilities or stimulating the development of new technology to meet them. As profound in their impact on all carriers as technological innovation are the secondary consequences of both interconnection competition and SCC entry that ensue from the jurisdictional separations process.

In addition to revenue decreases ensuing directly from losing a piece of terminal equipment to a competitor, there is the indirect loss of the share of the interstate revenues pot claimed through the separations process. On the terms illustrated above in Figure 1–27, a $1,000 decrease in local plant investment reduces the allocation to interstate plant by $240, hence the annual revenue from interstate settlements by at least $72. Nonetheless, contrary to the position of other carriers, United has taken the position that "the terminal equipment market should no longer be viewed as a monopoly" [96] provided that the FCC and state regulators permit it to compete in that market on a *nonregulated* basis according to yet unwritten rules and that reasonable measures to prevent harm to the network are devised and implemented.

What the carriers fear is "regulated competition," meaning government sanctioned cartelization that ties their hands and protects "infant industries" from the established carriers and from one another. Conversely, however, the infants know they were born only because the regulators said "let there be competition": Established carriers see the FCC's restrictions on DDS (Dataphone Digital Service) pricing as exemplifying how the FCC, having taken it upon itself to create a new competitive infant industry, has also undertaken to protect the fledgling, thereby apparently stretching its interpretation of its own declared policy "to promote and maintain an environment within which existing and any new carriers shall have an opportunity to compete fully and fairly in the sale of specialized services" [97].

In mid 1975, however, the chief of the FCC's Common Carrier Bureau saw only minimal impact by new competitors on the prices and revenues of established common carriers [98]. The interim decision on Hi-Lo issued in September 1975 pointed to private line interstate revenues for 1974 that exceeded AT&T's "projections of revenues without the SCC competition" [99]. At the other extreme, a study issued in January 1976 for the U.S. Independent Telephone Association estimated a 73 percent average annual growth rate for the SCCs' market share between 1975 and a 10 percent market share in 1985 [100]. But all this addresses only the superficial effects of competition, not the massive warfare at the core over the very structure of the telecommunications industry. The structural questions, however, are not lost on the Antitrust Division of the Department of Justice, which filed suit against AT&T in 1974, thereby raising the possibility of major alteration of the structures that provide information services in the United States.

As we detailed in Section 5, existing separations procedures reflect years of painstaking political adjustments among the carriers, the regulators, diverse service offerings, and their clienteles. As also noted in Section 5, any diversion of revenues from interstate MTS and WATS to interstate private line services within or outside AT&T diminishes intrastate revenues from separations, hence threatens the intrastate local and toll rate structures. Whatever arguments may be offered in terms of the virtues of competition and of growing returns for all in a growing market, SCC capture of any share of the private line market—primarily big business customers—by leave of the federal regulatory authority threatens not only the revenues of the established carriers but also the very foundations of their political accommodations with the regulators and the customers/voters in fifty states.

In the past, the telephone companies have found it advantageous, from time to time, to shift revenues and costs according to changing patterns of demand for local and toll services and according to shifting patterns of state and federal regulation. AT&T originally opposed the station-to-station separations principle but, according to Richard Gabel, by the early forties:

> The Bell System . . . recognized the power of separations to offset latent pressure for interstate rate reductions. Acceptance of the station-to-station principle by the Bell System may be considered another landmark in separations history. With this step, the carrier officially recognized the dual character of separations principles; first, its formal content of serving as a means of separating property, revenues, and expenses between jurisdictions and, second, as a device to maneuver price and revenue requirements. *The formal content became subordinated, in the course of time, to its political function of capitalizing on the competitive regulatory interests of state vs. federal regulatory bodies.* [101] [emphasis added].

By 1975, the FCC-decreed competition by SCC's and interconnection equipment suppliers had led AT&T to meet competition in private line service and loss of local service revenues to interconnection equipment suppliers by moving toward cost-based pricing policies. Various ongoing tariff proceedings, Hi-Lo among them, reflected cost allocations or contributions to the common cost pool tailored to meet price competition where it occurred. In the process, whether through direct prices increases in other services or through indirect paths such as the separations changes induced by shifts from WATS or MTS to PLS, the tendency is toward making waves in services used not only by news media but by nearly every man, woman, and child in America.

The effect on actual revenues is ambiguous. If one accepts the idea that local demand is inelastic, revenues from that source might actually rise if prices increase; with more elastic demand for MTS, revenues and separations settlements might decrease; and the net result might therefore be either a loss or a gain for the established carriers. However, the political consequences of petitions for rate increases in the local services and in intrastate or interstate MTS are unmistakably clear. Competitive regulatory interests with distinct constituencies are thus a major factor in the telecommunication arena, whether the regulated are playing these interests off against each other or being ground between their opposing millstones.

In general, state regulators have faced pressure to keep local rates low, and to bring state toll rates in line with the lower interstate toll rates. Even as late as 1970, separations procedures were seen as the way to fend off higher state rates, even if that meant taking on the FCC. As described by Commissioner Johnson in dissenting to the FCC's adoption of the Ozark Plan:

> In late 1969, after a round of continuous surveillance proceedings, the Bell System agreed to reduce interstate rates. The state regulatory commissions, faced with rising costs for state telephone services and incessant petitions for rate increases by Bell's operating companies, were incensed. How could the FCC reduce rates when the states were being forced to raise their rates? The solution was obvious—rather than reduce interstate rates, the FCC should allow a change in the separations formula and shift some costs from intrastate to interstate, soaking up the excess profits in the interstate segment and providing relief for the regulatory commissions caught between consumers and the Bell System . . .
>
> But in January 1970, the FCC was not playing the game. Stung by past criticism that the FCC had simply acquiesced in the use of the separations procedures to accomplish political ends, the FCC refused to stop the rate reductions and they went into effect. Since the FCC had embodied the separations procedures into agency rules, their manipulations [sic] was more difficult and required more time. . . .

As Commissioner Johnson says it, the impasse was clearly by a lifted congressional eyebrow:

> Legislation was introduced in Congress that would have removed separations matters from the jurisdiction of the FCC. In response the FCC on May 20, 1970, constituted a "Joint Board" . . . to consider separations changes [102].

But, by 1975, some states had got the message: the New York PSC staff thought it "highly doubtful that the FCC would be willing, through a change in separations procedures, to take on additional interstate costs" [103].

Plausible as these political explanations of the Ozark Plan sound, they are not the only conceivable explanations. Commissioner Johnson overstates the political case beyond realism in arguing that:

> ... any number of alternative allocation schemes would be of equal intellectual rigor. The mathematical formulae are only so much obfuscation and diversion. The only question that makes a hill of beans is what public policy consequences will *result* from choosing among the alternatives. And yet *that* question the majority ignores entirely in this proceeding [104].

After all, at some point, the most exquisite obfuscations and diversions might just become transparent to judges like those who noted in *Smith* v. *Illinois*, the 1930 separations case, that "extreme nicety is not required," but who nonetheless found that "it is quite another matter to ignore altogether the actual uses to which the property is put" [105], a logic which experts such as economists, fond of relating prices however vaguely to costs, and technologists, hooked on actual uses, might also find congenial.

Another line of argument would account for the thrust of the Ozark Plan through primarily technological and financial factors. Although an economic thesis known as the Averch-Johnson effect holds that "regulation in the United States systematically biases management decision of regulated firms in the direction of investing more heavily in plant and equipment than ... if they were unregulated" [106], this thesis has yet to be clearly demonstrated. Contrary conjectures are equally plausible. Suppose that in a period of tight capital, the interstate network were operating close to saturation. If interstate toll prices were to go on dropping, an elastic demand would therefore increase at a disproportionately higher rate; the quality of toll service might then degrade abruptly because of the way a hierarchical network (see Fig. 1–13) behaves when overloaded. Normally, when a direct route is unavailable, the network will seek and find an alternate route via a switching center higher up in the hierarchy. When the network gets overloaded, more and more of its resources go into fruitless searches for alternate paths, thereby further increasing the load and eventually blocking the entire network to any new toll calls. Under such conditions, a critical technical need to keep interstate prices from falling too rapidly would coincide with the desire to transfer costs from state services to interstate services. The observable phenomena described by Richard Gabel and Nicholas

Johnson would occur just the same, but for reasons significantly different from those they postulate.

But times changed after 1970, and in an inflationary period the FCC was confronted not with prospects for lower MTS prices, but rather with requests for MTS price increases. There is continuing pressure on the state agencies to find an "out" for local service costs rising because of inflation and other factors, including any effects of competition in the private line and interconnection markets. The National Association of (State) Regulatory Utility Commissioners has repeatedly challenged the federal thrust toward competition and supported a "Home Telephone Act" aimed at preserving present price structures. And the North Carolina Public Utilities Commission has challenged the application of the FCC's Carterfone decision to intrastate interconnection facilities; the case was first argued before the U.S. Court of Appeals (4th Circuit) in September 1975 and decided against North Carolina in April 1976 [107]. Since the relationship between federal and state jurisdiction is at issue, appeal to the Supreme Court is not unlikely. If the North Carolina PUC eventually wins, it will have effectively blocked any competition in interconnection and, in so serving its constituency according to its lights, it will have preserved for the telephone companies in its jurisdiction a monopoly in the terminal equipment market along with the corresponding separations revenues.

Similarly, in early 1975, contemplating the possible consequences of eliminating terminal equipment costs from the separations procedures, the staff of the New York State Public Service Commission gloomily envisaged that it:

> . . . would require an unprecedented increase in the interstate assignment of the remaining non-terminal equipment costs in order to preserve the current interstate cost assignment. A change of this magnitude would most likely require the abandonment of the court-tested relative use principle underlying the current procedures [108].

Which means that the thread tying evolving politics to the judicially sanctioned relative use criterion might snap after 45 years of stretching by ingenious devices like the "deterrent effects" described in Section 5.

AT&T has emphatically expressed its stand on the injection of competition in some telephone service markets, as to both principle and detail. As John de Butts, AT&T's board chairman put it in September 1973:

> ... the prospect confronting us is not free and open competition as an alternative to monopoly but rather a third alternative affording the virtues of neither and the disadvantages of both, and that is regulated competition—a division of the market place arbitrarily imposed and artificially maintained [109].

In the controversy over resale and shared use of carriers' private line facilities [110], AT&T tags the other parties as "that subsector of the public which have relatively large business communications needs" and which "have a natural tendency to equate the 'public good' to their own interests" [111]. These parties' arguments are dismissed as the "promotion of their own pocketbook interest in obtaining for themselves ... lower cost communications even though the result is that others [the residential user and the small businessman] will have to pay more" [112]. As we noted in Section 5, the White House Office of Telecommunications Policy voiced a contrary concern that overpricing of business services would be passed on to consumers as a hidden and regressive sales tax.

But some of those very parties already were long standing beneficiaries of exceptions to shared use prohibitions embodied in AT&T tariffs themselves (Fig. 1–46). These exceptions, like rock strata in a cliff, record the deposits of years of political accommodations among the regulators, the regulated, and their constituencies. Others among the parties understandably wanted in, either directly or through lower rates induced by competition (see Fig. 1–7). Thus, in spite of apparent contradiction, MCI could state that "because of the built-in system of subsidies which the Bell System supplies to its large business and TV private line customers, the average small business and residential user of telephone service has been required by Bell to pay substantially higher rates than would have been required in the absence of such subsidy" [113]. But then MCI itself aims mainly at the big business market.

There is much latitude for discretion left in the statute that guides the FCC and the courts in sorting through conflicting expert testimony and the contradictory statements that interested parties make in formal proceedings. On the one hand, there is the emphatic statement that:

> ... it shall be unlawful for any common carrier to make any unjust or unreasonable discrimination ... in connection with like communication service ... or to make or give any undue or unreasonable preference or advantage to any particular [party] ... or to subject any particular [party] to any undue or unreasonable prejudice or disadvantage [114].

But, says the statute, services "may be classified into day, night, repeated, unrepeated, letter, *commercial*, *press*, *government*, and such other classes as the Commission may decide to be just and reasonable, and different charges may be made for the different classes of communication" [115] (emphasis added).

Thus, when the Associated Press once sued the FCC, alleging unlawful overcharges, the court could hold that:

> . . . not every variation in prices charged customers for a particular feature of the carrier's service supports a claim of unlawful discrimination. Since rate classification, no less than other classifications, may be justified by differences between the classes, the mere existence of a disparity between particular rates does not establish a statutory violation. Much less can an allegation of discrimination safely be predicated upon a piecemeal comparison of classified tariff charges [116].

Hence the pervasive and persistent alchemist's urge to transmute the inherently arbitrary and judgmental patterns outlined in the preceding sections into true fact or valid theory. Hence bizarre but intense quasi-judicial and judicial nit pickings over what are "like" or "unlike" services or over what is "specialized" or "ordinary." Hence the drive to assert definitively the presence or absence of discriminatory patterns or the likelihood that change will be discriminatory (Fig. 1–47).

But the touchstone of it all is consensus within the polity as to what is "just and reasonable." So, in the spring of 1976, AT&T and its allies put the polity to the test. A series of nearly identical bills, all entitled "Consumer Communication Reform Act of 1976," were introduced in both houses of Congress "to reaffirm the intent of Congress with respect to the structure of the common carrier telecommunications industry rendering service in interstate and foreign commerce" [117]. The findings and declaration of purpose and the specific provisions of all these bills set forth the essence of AT&T's case for the Restoration. At least one of the House versions would authorize AT&T to acquire its competitors, a provision absent from the Senate version. By turning the clock back to the early sixties or before, the bill would restore the telecommunications industry structure to what it was when competition challenged monopoly mostly in theory.

(Text continues on page 137 . . .)

Manufacturer's Share of Interconnect Market Lines of PABX Switching Equipment, Year End 1970-1972

Manufacturer	1970 Systems	1970 Lines	1971 Systems	1971 Lines	1972 Systems	1972 Lines	1970-1972 Cumulative Systems	1970-1972 Cumulative Lines	% Total
Hitachi	320	65,000	270	55,000	290	60,000	880	180,000	31.5
OKI Electronics	100	8,000	350	34,000	330	72,000	780	114,000	19.9
L.M. Ericsson	75	5,000	370	26,000	420	35,000	865	66,000	11.5
Nippon Electric	90	8,000	160	17,000	300	35,000	550	60,000	10.5
Fujitsu	60	5,100	115	10,000	200	30,000	375	45,100	7.9
ITT	55	4,500	80	11,000	100	12,000	235	27,500	4.8
North Electric	50	5,000	50	5,000	50	10,000	150	20,000	3.5
Plessey	–	–	50	4,000	110	10,000	160	14,000	2.4
Norelco (Philips)	25	2,700	35	3,000	100	7,000	160	12,500	2.1
Stromberg-Carlson	–	–	–	–	12	4,000	12	4,000	.70
Tele/Resources	–	–	–	–	20	1,500	20	1,500	.26
Northern	–	–	–	–	10	800	10	800	.14
Other	55	4,000	90	6,000	150	15,000	295	25,000	4.3
TOTALS	830	107,000	1,120	171,000	2,092	292,300	4,492	570,400	

NOTES: Above figures represent manufacturer *shipments* of lines of switching equipment, not installed systems.

(a)

Industry Penetration

• Hotel/motel industry	40%
• Manufacturing industries	15%
• Wholesale/retail industries	10%
• Hospitals	5%
• Banks, brokers, and financial institutions	5%
• Business service organizations	5%
• Automobile dealers	5%
• Other industries	15%
TOTAL	100%

(b)

Figure 1–45. Stakes in the Interconnection Market

The private automatic branch exchange (PABX) market is among those that went competitive following the 1968 Carterfone decision. An average hotel PABX handles 150 individual telephone lines. In 1972, some 300,000 lines were shipped by competitors, or 17 percent of the 1.8 million lines that Western Electric is estimated to have shipped that year. The international implications of the changing market are evident.

Source: Dittberner Associates, Inc. (Washington, D.C.), *Interconnection: An Economic Impact Analysis*, Tables III–1 and III–2, p. 45.

AT&T Tariff No. 260 Private Line Service

2.2 Use

2.2.1 A private line service may be used for one or more of the purposes specified in (A) through (J) following. When a private line service is arranged for joint use the joint user shall be permitted to use such service in the same manner as the customer as specified in (A) through (I) following.

(A) For the transmission of communications to or from the customer and relating directly to the customer's business.

(B) For the transmission, to all stations simultaneously, of communications which relate directly to matters of common interest to the customer and the authorized users when those connected to the service are all in the same general line of business.

(C) For the transmission of communications relating directly to the business of a subsidiary corporation over which the customer exercises control through the ownership of more than 50% of the voting stock.

(D) For the transmission of communications to or from any station on a service furnished to a Department or Agency of the United States Government when the head of the Department or Agency, or his duly authorized representative, notifies the Telephone Company in writing that the use is intended only for official United States Government business.

(E) Where the customer is an organized stock or commodity exchange, for the transmission of communications to or from an exchange member located on the floor of such exchange and relating directly to the business of the member.

(F) Where the use of the service relates to coordination or exchange of electrical pooled power, for the transmission of communications between any two or more stations of such service or similar services furnished to others who are parties to the coordinating or exchange arrangement.

(G) For the transmission of communications to, from, within and between air carriers, where the customer is an aeronautical communications company licensed under the Aviation Services rules of the Commission to operate stations in the aeronautical mobile and fixed services.

(H) For the transmission of communications to or from any station on a service furnished to the United States Postal Service for its use in the provision of its Facsimile Mail Service.

(I) For the transmission of communications of a state or local government agency where the service is ordered for such agency by the United States Government pursuant to the Intergovernmental Cooperation Act of 1968.

(J) Where the customer is a composite date service vendor, for the transmission of switched data (non-voice) communications for its patrons when such communications relate directly to the business of such patrons, for communications over the private line service in accordance with (A) preceeding, and, additionally, for composite data service or hybrid data processing service, as defined in Part 64.702 of the Federal Communications Commission's Rules and Regulations, rendered by a patron of the customer. All other forms of collection, transmission or delivery of communications for others is prohibited.

Figure 1–46. Exceptions to the Prohibition of Shared Use of Private Line Service

Airlines, stock brokers, power companies, and various levels of government are among those authorized to make joint use of common private line facilities.

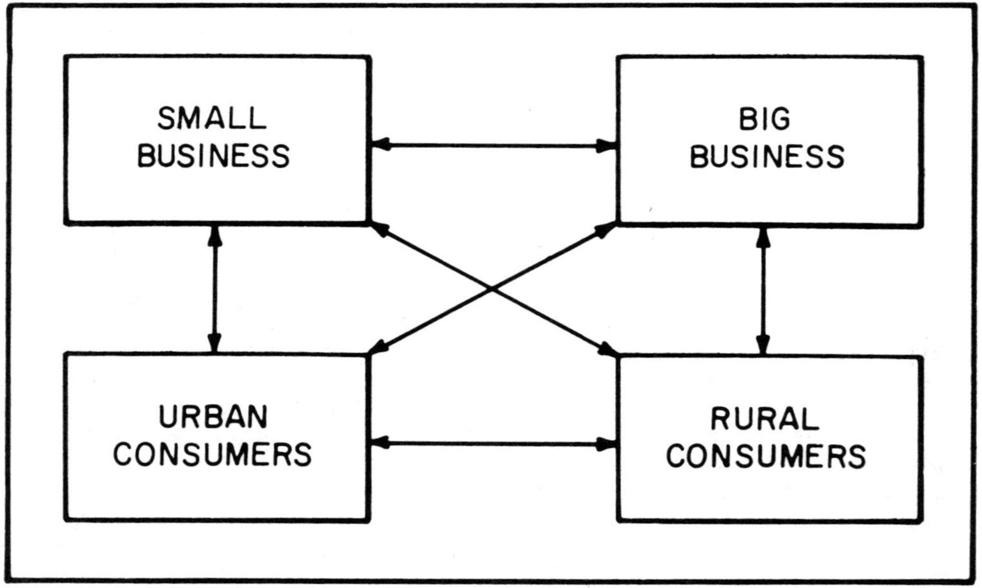

Figure 1–47. Equities and Cross-Subsidies

Who benefits at whose expense? Whose benefits are increased or reduced by a proposed change? For example, the data of Figure 1–30a support the idea that urban consumers help pay for service to rural consumers and business services help pay for residential services. The use of private lines by big businesses, the insensitivity of private line charges to the extent of actual use, and the sharing exceptions of Figure 1–46 suggest that small businesses and residences help pay for big businesses. The jurisdictional separations process benefits intrastate users at the expense of interstate toll users. But what is the overall net effect on each party? What, indeed, are appropriate definitions of the parties? On what basis are costs to be measured?

9. WHEN AND HOW—A PUBLIC POLICY AGENDA

Competition vs. monopoly. In the American ethos, this stark opposition implies the preferred industrial structure and the adoption of policies that support it. But the case for either extreme is not as clear cut as advocates would have it. New entrants tend to favor open competition while entering and protection from further entrants once in. Nor are these polar forms of industry structure as self-evidently desirable ends to everyone in the United States or abroad as they are to latter day Adam Smiths and to postal, telephone, and telegraph managements.

Modern economic theory recognizes that various goals—such as profit maximization or various shapes of income and wealth distribution—and various conditions of production—such as the extent of economies of scale—are best accommodated by various shadings of the extreme structures. Both the regulators and the regulated monopolies have espoused the goal of "service to all" and have justified monopolistic averaging and value-of-service pricing policies in the name of that goal. But despite all the talk, service to all did not in fact come at all close to materializing without both a postwar boom in the economy and the explicit government subsidies provided through the Rural Electrification Administration and the Rural Telephone Bank.

Throughout the early seventies the White House Office of Telecommunications Policy was for competition in the name of efficiencies of the invisible hand as contrasted to the inefficiencies of hidden subsidies, but even it conceded that "economic support for certain classes of telephone users is [not] necessarily bad policy." Finally, as suggested by Figure 1–48, structural policy is far from being the sole form of governmental influence on economic processes.

Thus the antinomies stemming from opposing monopoly to competition mask real political differences over important ends with specious distinctions between partial means. What are the ends? "To make available, so far as possible, to all the people of the United States a rapid, efficient, Nationwide and worldwide wire and radio communication service with adequate facilities at reasonable charges" [118] is the overarching purpose Congress laid on the Federal Communications Commission and receptive regulated carriers by the Communications Act of 1934.

At that time, 31 percent of households had telephones and 6 million business telephones were in service. By 1974, there were telephones in 94 percent of households and close to 39 million business telephones were in service. In 1934, the leaps in electronics and electronics industries stimulated by World War II still lay ahead. There were neither the undersea telephone cables nor the satellites that now make worldwide service a real possibility as well as hopeful rhetoric. The strategic role of telecommunications—economic as well as military—was far less significant than it is today. A service was then virtually indistinguishable from the technologies, the facilities, and the institutions used to provide it and one such combination was clearly distinct from others. There were no electronic computers, hence neither compunications nor the institutional boundary marking issues that this merged technology raises. In 1934 there was scarcity and specialization of communication services; today there is versatility and abundance of information resources of many kinds.

Defining Policy Aims

The first order of business is therefore to examine the validity of the mission set forth in the Act of 1934 by asking the following questions.

1. What should be our policy aims for the 1980s and beyond? The answer suggested by the legislative proposals of early 1976 is that future aims should be precisely what they were in an already distant past. It is true that most telecommunications is still from mouth to ear. But the evidence in preceding sections tells us that enough else is already here, growing, and important enough to enough publics—*the* public among them—to deserve better than plans to fight the last war over again. At the very least, *what* is to be made available to all the people of the United States—and *who* pays for it—needs to be defined in keeping with the times. Hence the next question.

2. What is to be the scope of policy for the 1980s and beyond? Traditional institutional and technological distinctions are inadequate to deal with this question when AT&T and IBM, to name only the biggest traditional communications and computer companies, play on the same turf with the identical toys of compunications science and technology, and when the framework of the Act of 1934 has already been stretched to the limit, as for instance by the FCC and the Supreme Court wrestling to fit CATV in the crack between Title II, governing common carriers, and Title III, governing broadcasters. If "wire and radio communication service" is too narrow, and all of information resources too broad a reach, then suitable intermediate frames must be considered.

3. **Who is to decide what the aims and scope of policy should be?** Although the FCC has strained its congressional bounds mightily in recent years [119], there are fences even it has chosen not to climb and paths that the courts have barred to it. Hence, even assuming it were a desirable agent of change—and this is far from obvious—it is one of limited jurisdiction. And, as an agency in the Executive Office of the President, the Office of Telecommunications Policy will reflect at most the policies of a given Administration. The intervention of broad public debate and of congressional action are both timely and essential, not only to define the aims and the scope of policy for the 1980s and beyond but also to shape appropriate instruments to carry out the chosen policy.

4. **What are the limitations of present policy instruments?** Given that present policies provide shared federal and state authority over common carriers, confused authority over CATV, and sole federal authority over broadcasting, how consistent is this jurisdictional pattern with aims for the future? If the subsidies provided by urban telephone users to their rural counterparts are an artifact of older patterns of legislative representation, how consistent is this with shifts induced by the Supreme Court's decisions in the reapportionment cases? [120] How consistent with trends in tax policy? Granted that past processes have favored some suppliers over others and some consumers over others, how consistent are the results with aims for the future? If they prove to be inconsistent, how might any losers be fairly compensated for contributing to the greater future good?

Regulatory agencies, long favorite instruments of policy in diverse fields, are under renewed and widespread attacks. The lessons that John Dunlop drew from a stint as Secretary of Labor are instructive. The regulatory approach, says Dunlop, leads to the following problems:

> First, it encourages simplistic thinking about complicated issues.
>
> Second, oftentimes policies that appear straightforward will have unintended consequences which can create problems as severe as those with which the regulations were intended to deal.
>
> Third, the rulemaking and adjudicatory procedures of regulatory agencies tend to be very slow, creating conflicts between the different groups involved, and leading to weak and ineffective remedies for the people the programs aimed to help.
>
> Fourth, the rulemaking and adjudicatory procedures do not include a mechanism for the development of mutual accommodation among the conflicting interests.
>
> Fifth, a further problem is what is called "regulatory overlap," where a

number of different regulatory agencies share some of the same responsibilities.

Sixth, regulatory efforts are rarely abandoned even after their purpose has been served [121].

There are masses of hopeful recipes for regulatory reform (Fig. 1–49). There are also gloomier views of the reform of American administrative law, such as Richard Stewart's conclusion that "given [in Hegel's words] 'the undefined foreboding of something unknown', we can know only that we must spurn superficial analysis and simplistic remedies, girding ourselves to shoulder for the indefinite future, the intellectual and social burdens of a dense complexity" [122]. But all these writings address themselves to the same generic instrument, and fail to ask, as suggested by Figure 1–48, the next question.

5. **What policy instruments—institutions and processes—might best serve our aims for the 1980s and beyond?** Contrasting competition to monopoly on economic (or ideological) grounds addresses only one criterion for choosing among alternative industry structures. Neglecting organizational, managerial and engineering criteria exposes policy to serious risks. Granted a vital interest in an effective nationwide telecommunications system—narrowly or broadly defined—is it safe to continue entrusting it mainly to one leading organization that conceivably might, some day, jump, fall, or be pushed the way of the Penn Central Railroad? On the other hand, could this system be reliably and economically operated and updated in fragments, by competing organizations, with competing suppliers and competing repairmen all pointing the finger at one another as the cause of any trouble?

Granted—and this is not universally accepted today—that technological innovation is of continuing high importance, do the pressures generated by state, local and federal regulation and cost allocations adequately guide management decisions as to how much is spent for the development and implementation of which new communications technologies by Bell Laboratories and others? Would more overt political control, such as with the National Aeronautical and Space Administration, provide better response to public demands? Or would replacing corporate and administrative decisions by overt political control make basic research and development suffer at the expense of some short term social goal or program perceived to be more important by the fashion of the moment? Would greater market competition stimulate innovation or so fragment the means as to stifle it?

Granted that what is on the inside and what is on the outside of compunications facilities is increasingly hard to tell. Granted that the relationship between facilities and services is increasingly complex. And granted that a consolidated ATTIBMITTGERCA worldwide compunications facilities and service system would likely be unmanageable—even if it were ideologically acceptable—what would an optimal structure look like? What are the most important yardsticks, among equity, cost, responsiveness, reliability, accessibility, confidentiality, and many others, by which optimality both should and could be measured? Would the present court-and-commission sanctioned placement of the computer/communications boundary, or of the interconnect and SCC competition, be among the optimal cuts? Could it lead toward any of them?

6. How best to match up industry structure with performance aims? In a world of increasing interdependence, domestic and international affairs are more tightly meshed than ever before. With global interdependence, of which compunications is the very nerve system, control over these facilities and services through government policies is a powerful lever on domestic and international affairs alike. And so, the final question.

7. How best to harmonize domestic and foreign policy instruments?

- Laissez-faire
- Taxes
- Tax write-offs
- Regulation of price, quality and entry
- Standards
- Research and development
- Prohibitions; financial and criminal sanctions
- Subsidies
- Rewards for innovation and invention
- Incentives, e.g., matching funds
- Model legislation
- Import/export trade management
- Information exchange
- Establishing or supporting an industrial base by government purchase
- Establishing new institutions
- Government control or monopoly
- Building civil works
- Propaganda
- Fear

Figure 1–48. Policy Instruments

Although attention to government regulation has tended to focus on such means as the copyright laws or the Federal Communications Commission, the range of the tools of government that may or may not be exercised is far wider. (Listing adapted from Joseph F. Coates, *Structural Failure: The Case of Local Government*, an unpublished paper.)

Proposals to Reform the FCC*

1934 American Bar Association Committee on Administrative Law
- abolish independent regulatory commissions and shift all but their specifically judicial work to Executive Branch bureaus
- place judicial functions in an administrative court

1936 E. Pendleton Herring, Federal Commissioners
- appoint more people to commission with industry experience

1937 President Roosevelt's Committee on Administrative Management
- define FCC decision-making standards more clearly
- shift administrative tasks of commissions to Executive Branch departments and make these departments responsible to a bureau headed by one person

1941 Attorney General's Committee on Administrative Procedure
- clearly delegate functions and authority within each agency
- establish a separate unit of hearing officers within each commission and provide for appeals of their decisions to the commission chairman

(this report provided the basis for the Administrative Procedure Act of 1946)

1941 Robert Cushman, Independent Regulatory Commissions
- separate executive from judicial functions in commissions, assigining executive work to Executive Branch departments and judicial work to an administrative court.

1949 The First Hoover Commission
- assign executive functions that are not specifically "regulatory" to the Cabinets
- FCC staff should be organized along functional (broadcasting, common carrier, etc.) rather than professional (legal, technical, accounting, etc.) lines

(a 1952 amendment to the Communications Act reorganized part of the FCC staff into functional bureaus)

1951 President Truman's Communication Policy Board
- make the Commission's policy standards more specific
- establish a long-range planning capacity at the FCC

1955 The Second Hoover Commission
- remove judicial functions from commissions and place these functions within courts of general jurisdiction and special administrative courts

1959 Ronald Coase, "The Federal Communications Commission"
- Congress should direct the FCC to auction off broadcasting licenses to the highest bidder

1959 Bernard Schwartz, The Professor and the Commissions
- insulate the FCC from Executive Branch control by allowing the FCC to submit its budget directly to Congress (rather than through the Bureau of the Budget, now the Office of Management and Budget) and prohibit the President from removing an FCC Chairman

1960 Landis Committee Report to President-Elect Kennedy
- extend FCC Commissioner tenure to ten years
- authorize White House staff members to oversee regulatory agencies
- create an administrative court for commissions

1962 Henry Friendly, The Federal Administrative Agencies
- each decade an appropriate Congressional subcommittee should review all laws applying to a particular commission
- create more specific standards for the FCC than those contained in the Communications Act

1962 Booz, Allen, and Hamilton, Organization and Management of the FCC
- the FCC needs more money and staff, but most importantly a better definition of standards

1964 Newton Minow, Equal Time
- assign judicial functions of the FCC to an administrative court
- assign executive functions of the FCC to a single administrator appointed by the President

1969 President Johnson's Task Force on Telecommunications
- create an Executive Branch agency and assign it general responsibility for allocating the radio spectrum
- the FCC should continue to license operators within spectrum areas assigned by this Executive agency

Fig. 1–49. cont'd overleaf

*Included are proposals that apply to all independent commissions as well as to the FCC.

Fig. 1–49. cont.

1970 Nicholas Johnson, How to Talk Back to Your Television Set
- -strengthen the organization of consumer interests by establishing a Citizen's Commission on Communications with the authority to appear as an advocate for the public interest before the FCC

1971 Walter Emery, Broadcasting and Government
- -lengthen the tenure of FCC Commissioners to ten or twelve years
- -increase the FCC's information-gathering capacity

1971 President Nixon's Adviso y Council on Executive Organization
- -reduce the size of the FCC from seven to five Commissioners
- -the FCC is the only independent commission that should retain the collegial structure to insure broad-based deliberation and a non-partisan environment

1971 Brookings Institution Report on Reforming Regulation
- -possible options reported in this work:
- -independence with Presidential oversight
  - -authorize the President to make general rules and policies for commissions through the use of executive orders
- -a regulatory czar
  - -adopt Rexford Tugwell's New Deal proposal to create a second Vice President responsible for government regulation
- -legislative reform
  - -eliminate the formal procedural requirements of the Administrative Procedure Act
- -constituent representative
  - -assign each position on a regulatory commission to a representative of a particular interest

1972 Sidney W. Head, Broadcasting in America
- -the FCC should develop better standards, particularly for station licensing and renewals

1973 Erwin G. Krasnow and Harry M. Shooshan, "Congressional Oversight"
- -increase the Congressional oversight capacity by providing more staff and funding to the committees responsible for reviewing the FCC's work
- -increase the staff and funding of the FCC
- -development by Congress of more meaningful standards and guidelines for agency performance

1973 Roger G. Noll, Merton J. Peck, and John J. McGowan, Economic Aspects of Television Regulation
- -the FCC needs a new legislative mandate specifying the following:
- -broadcasters should be given the same first amendment freedoms regarding program content and quality as the print media
- -competition should replace localism as the primary broadcasting standard
- -the FCC's broadcast licensing concern should be limited to engineering and other technical aspects of spectrum management

1973 Rolla E. Park (ed.), The Role of Analysis in Regulatory Decisionmaking
- -the FCC needs to increase its analysis and planning capability

1974 Henry Geller, "A Modest Proposal to Reform the Federal Communications Commission."
- -decrease the number of FCC Commissioners to five
- -extend Commissioners' tenure to fifteen years and bar reappointment
- -do not allow Commissioners to work for the Communications industry until ten years after their tenure with the FCC is completed
- -increase the quality of FCC Commissioners
- -eliminate hearing procedures when facts are udisputed and use written submissions
- -as suggested by former FCC Chairman Dean Burch:
- -create two divisions at the FCC along these lines:
  - -broadcast/cable
  - -common carrier/safety and special services
- -assign each division to half of the Commission
- -the Chair would be responsible for convening the entire Commission on issues that overlap the divisions

. . . continued →

Fig. 1–49. cont.

1975	Committee for Economic Development, Broadcasting and Cable Television -confer FCC adjudicatory functions to a new communications court following the pattern of the U. S. Tax Court when it was part of the Executive Branch -the President should appoint judges for twelve year terms with Senate approval -Congress and the FCC should strengthen the Office of Plans and Policy particularly by staffing the Office with more economists, engineers, attorneys, and social scientists
1975	Congressman Torbert Macdonald's Bill to Reform the FCC -reduce the Commission size from seven to five -increase the tenure of Commissioners from seven to ten years -double the amount of staff assistance for each Commissioner -require Commissioners to keep a record of contacts with lobbying groups -make all meetings public unless it is determined by a majority vote of the Commission that a public meeting would be contrary to the public interest -have the FCC go directly to Congress for appropriations rather than through the Office of Management and Budget -make the President's designation of a Chairman subject to Senate approval -have the FCC provide Congress with immediate access to documents and publish a quarterly rather than an annual report

Figure 1–49. Proposals for Regulatory Reforms

Since the FCC was established by the Communications Act of 1934, numerous proposals have been made to reform it and other regulatory bodies.

Source: Vincent Mosco, *Reforming Regulation: The FCC and Innovations in the Broadcasting Market*, Program on Information Technologies and Public Policy, Harvard University, Cambridge, Mass., Publication No. 76–2, Figure 1, p. 2–3.

New phone rates filed. Here's how you could be affected.

Some highlights:

- **Monthly charge for service goes up 34¢ for residence, 57¢ for flat rate business, and $1.04 for message rate business.**
- **No timing of one message unit calls.**
- **Message units cheaper between 9PM–9AM weekdays, all day on weekends and 5 major holidays in New York and Buffalo Metropolitan Areas.**
- **Overtime charging on all multi-message unit calls starts after 3 minutes.**
- **Instate long distance rates up; new 2-minute rate for long distance calls you dial yourself.**
- **Surcharges for person-to-person and operator-handled local and long distance calls increase.**
- **10¢ pay phone calls stay at a dime.**
- **Minimum service connection charges up: to $18.50 for residence, to $44 for business.**
- **Residence extensions increase 34¢.**
- **Business extensions and supplemental services increase.**

New rates for telephone service, designed to produce $297.9 million in added revenues annually, have been filed, subject to Public Service Commission approval.

The filing follows a Commission order of October 22, after 11 months of hearings and review, granting a telephone rate increase in that amount.

The increase is to help us meet rising taxes, wages, cost of modernization and replacement of facilities, and other costs of providing telephone service.

Following are some of the changes in more detail (rates quoted do not include taxes):

Basic service rates

The monthly charge for service increases 34¢ for residence customers, 57¢ for flat rate business customers, and $1.04 for message rate business customers. PBX trunk rates remain unchanged.

For some customers in the Mid-Hudson and Niagara Regions, Buffalo Suburbs, Utica, Esperance, Galway, Sackets Harbor, Silver Creek, Attica, Varysburg and Westerlo, the increase in basic rates is offset by changes in the basis for establishing monthly charges. In almost all cases, this results in net decreases ranging from 9¢ to $1.37 a month, depending on location.

Flat rate customers in 11 communities—Montauk Point, Kings Park, Bayport, Shoreham, Pleasantville, Brewster, Tarrytown, Ossining, New City, Ellenville and Saugerties—are receiving additional increases due to reclassification to higher rate groups. Customers affected are being notified by mail.

Message units

The one message unit call remains untimed—no additional message units apply for one unit calls lasting longer than 5 minutes.

The regular price of a message unit remains at 8.2¢, except as indicated below.

In New York City, Long Island, southern Westchester (including Ossining, Pleasantville and Armonk) and the Buffalo Metropolitan Area, a Message Unit Discount Plan drops the message unit price to 6¢ between 9PM and 9AM on weekdays, and all day on Saturdays, Sundays and 5 major holidays. That's a saving of 27% for message rate customers, and for Basic Budget and flat rate customers who make message unit calls. (For message rate customers, the saving applies after the monthly allowance of 50 message units is used).

Overtime charging on calls of two or more message units begins after 3 minutes—instead of 5 or 4—in New York City, southern Westchester, Nassau, Cold Spring Harbor, Farmingdale and Amityville. And most overtime charges are up.

If your initial message unit charge is 2 units for 3 minutes, the overtime charge is 1 unit for each additional 2 minutes. For 3, 4 or 5 message units, it's 1 unit for each additional minute. And for 6 or 7 units, overtime is 2 message units for each additional minute.

The Message Unit Discount Plan applies to multi-message unit calls as well as to one unit calls.

Your business office representative can help with any questions you may have about any message unit usage shown on your bill. For customers who would like a photostatic copy of their Message Unit Details, where available, it can be furnished at a cost of $1.50 for each request for residence; $1.50 plus 25¢ a page for business customers.

Basic Budget Service

Where available, Basic Budget Service for residence customers increases from $4.03 to $4.37 per month. As before, there is *no* message unit allowance; all local calls are charged at 8.2¢ (or 6¢ after 9PM weekdays, and all day on weekends and 5 holidays in the New York and Buffalo metropolitan areas).

Long distance

Overtime on direct dialed non-coin calls *within the state* between 8AM and 11PM begins after 2 minutes instead of 3. Calls of 2 minutes or less can cost up to 20% less.

Long distance rates are increased 17% overall. However, this increase is not applied evenly.

- Direct dialed non-coin calls placed during "Day" hours (8AM to 5PM weekdays) decrease up to 14% under 2 minutes, and increase up to 35% under 3 minutes. The percentage increase for calls up to 186 miles is slightly higher than that for calls of longer distance.
- The "Evening" discount (after 5PM weekdays and all day weekends) ranges from 20% to 48% below the new daytime rates. While evening calls of less than 2 minutes are as much as 20% less than in the past, calls of more than 2 minutes are up to 34% more. Generally, the percent increase is greater for calls of greater distance.
- Lowest one-minute "Night" rates (11PM-8AM all days) are 50% to 60% lower than new daytime rates for calls of the same duration. Short distance "Night" calls decrease up to 27% below present rates, and longer-distance calls increase up to 28%.

Surcharges on operator-handled long distance calls within the state increase from 30¢ to 60¢, except for person-to-person. On person-to-person, the surcharges increase from 50¢, 60¢ or 70¢ to $1, $1.20 or $1.40, depending on the distance called. Operator-handled long distance calls continue to have a 3-minute initial period. The surcharges are to be added to the direct distance dialed rate for 3 minutes.

Surcharges for operator-handled non-coin *local* calls increase to 60¢ except for person-to-person. On person-to-person, the surcharge is $1 for local calls. On coin calls, surcharges range from 40¢ to 60¢ for operator-handled calls, except for person-to-person, which range from 80¢ to $1.00.

Surcharge calls—both local and long distance—include those placed by the operator at customer request, and collect, credit card, third number, time and charges, special billing number calls, and person-to-person. A surcharge does not apply on normally dialed calls if you encounter trouble and *ask the operator for help*. Nor does it apply for handicapped persons requiring operator assistance on a call that is regularly dialed. Surcharges do apply for handicapped persons, however, on all collect, credit card, third number, time and charges and person-to-person calls.

There are *no* changes in rates for *out-of-state* long distance calls.

Coin telephones

The 10¢ local coin telephone call remains unchanged. However, *other local* coin calls to more distant points generally increase in amounts ranging from 5¢ to 20¢ for the initial period; in addition, the initial period becomes 3 minutes instead of 5, and some overtime rates increase. *Long distance* calls made during the day and evening from coin phones increase 3¢ to 27¢ above direct-dialed long distance rates from non-coin phones for the 3-minute initial period. The rates for long distance operator-handled and person-to-person coin calls are the same as for comparable non-coin calls. Night coin rates are the same as evening coin rates.

Service connection charges

The minimum connection charge increases from $15 to $18.50 for residence, and from $30 to $44 for business. Separate charges of $3.50 for residence and $22 for business apply to each additional one-key telephone connected—higher charges for key telephones.

Similar charges apply to moves, changes, or rearranging of phones, and to changes in classification of service; a lower charge applies when changes do not require a visit to the premises. *And there is no charge when a customer changes to a less expensive classification of service within 60 days of a rate increase.*

These new charges cover the expenses involved in issuing the order for service, performing central office line work, and for work done on a customer's premises. The amount of the charge depends on whether one or more of the above work operations is required.

Residence extensions and premium items

The price for residence extension phones increases from $1.34 a month to $1.68.

Touch-Tone® Service to increase from $2.54 a month to $3.05 for residence customers; $3.68 to $4.42 for business.

Trimline® phones to increase from $1.95 to $2.34 a month. Single Payment Option charge from $101.30 to $121.56.

Princess® phones to go from $1.38 to $1.66 a month. Single Payment Option charge from $50.65 to $60.78.

Major business service increases

- Extensions—flat rate goes up 57¢, to $2.85 a month; message rate goes up $1.04, to $2.85 a month.
- Centrex—line rate increases range from 17.5% to 25%.
- Key equipment—approximately 17%.
- Data and Teletypewriter equipment—20% to 25%.
- PBX systems—most increases range from 20% to 25%, depending on the type of equipment involved. A few increase up to 45%.
- Private line-mileage services—some, 25%; others, increased earlier this year, 10%.
- Protective connecting arrangements for customer-owned equipment—changes range from a reduction of $5.02 to an increase of $2.07, depending on the type of equipment involved.
- Instate WATS services—the average customer's bill will increase approximately 17%.

Your business office representative is available for further explanation of the rates filed.

New York Telephone

THE NEW YORK TIMES, FRIDAY, OCTOBER 31, 1975

Part Two

Computer or Communications? Allocation of Functions and the Role of the Federal Communications Commission

Paul J. Berman

I. INTRODUCTION

Control over communications services is a great source of power in our society. The lives of most citizens are greatly affected by decisions as to what communications services are to be provided, who is to provide them, how and to whom they are made available, and at what prices. Clearly, the relative prices and availabilities of communications services affect whether a person will telephone, telegraph, send a mailgram or a letter, or travel to another place at any given time in order to communicate to another person. Whether or not a television broadcaster can feasibly obtain adequate point-to-point communications facilities may determine what stories are sent from remote locations to television news headquarters to be part of the evening news. The practicality of transferring large quantities of information from one computer to another has an impact on who is on what computerized mailing list and whether one credit agency obtains another agency's files. Computer output transfer or distribution services may in fact set their prices to compete among themselves for the computer output business. This competition may then affect the price and availability of other communications and transportation services. Not only is communications itself a big business [1], but the price and availability of communications services—results of specific allocation decisions—have effects on the price and availability of a wide range of other goods and services throughout society.

Government has traditionally been concerned with information distribution systems (e.g., the Post Office) as important factors in the cohesion and smooth functioning of society. For example, Congress created the Federal Communications Commission in 1934 to regulate and allocate both point-to-point (like telephone) and broadcast communications services [2]. Since that time, the computer industry has grown up, and in 1974 computer and communications technologies are becoming increasingly indistinguishable and interchangeable. Entirely new communications related services and functions have developed that no one could have imagined when the FCC was created. In some instances, the FCC itself has decided whether or not it will regulate and allocate these newly developed services and functions. In this process, important power over the underlying information functions—important power over people—has moved by default or device from Congress to the administrative agency.

Today, the answer to the question of what is a point-to-point communications function or, more practically, what is the scope of FCC power over such functions, is anything but clear. Issues arise as to how much power the FCC is to have as it addresses technological developments and organizational conflicts accompanying the increased use of computers in information transfer and processing.

In order that these developments and conflicts and their interaction with allocation and regulation systems can be fully understood, a few terms to be referred to throughout this article must be defined. Two-way *point-to-point communications* services by electrical means have traditionally been supplied by a few *communications common carriers*: the telephone and telegraph companies. These organizations operate networks, or systems of electrical paths used for communications between individuals at specific points. The primary task in operating such a network is the *transmission* or *carriage* of signals from place to place along the paths of the network. This carriage is the primary business of the network operators and is one of the major interests of the FCC.

Point-to-point communications, however, encompass more than the mere carriage of electrical signals from place to place. Someone or something must convert the vocal or written message of the sender into electrical signals suitable for transmission and reconvert the signals into voice or writing for the receiver. Someone must choose the best method and pathway for transmitting the message and see to the orderly operation of the network. In the law and the regulations of the FCC, these additional operations are known as services *incidental to carriage* [3]. Since they are clearly so important to the major business of carrying the signals, incidental services provided by carriers are also regulated by the FCC.

To connect users of a network with each other, the individual paths of the network must be interconnected. This interconnection function, also incidental to carriage, is called *switching*. Two types of switching are traditionally recognized. In *circuit switching*, as historically associated with the telephone, a path is established between two points for the duration of the conversation or message. In *message switching*, historically used by telegraph systems, the connection between the two points is not simultaneous, and the message may be stored at intermediate points along the route before being forwarded toward its final destination. However, as computers and related techniques increasingly serve this interconnection function, the distinction between the two types of switching is fading.

Data processing is dealing with information. Looking up a telephone number is data processing; putting a message into Morse code is data processing. In contemporary terms, data processing usually includes activities like gathering, filing, storing, manipulating, or using information. With the advent of the computer, much of the work that was once done by armies of clerks with adding machines and filing cabinets can now be done automatically.

As it happens, a great deal of data processing is necessary for providing communications services to the public. Billing, switching, traffic monitoring, and the like can be controlled swiftly, effectively, and inexpensively by computer. Computers are also employed in multiplexing, a set of techniques that enables one wire to carry many conversations without getting them confused.

Thus, Bell Laboratories, American Telephone and Telegraph's research arm, does research about computers and their applications [4]. By the same token, computer entrepreneurs continue to incorporate communications facilities into their computer service offerings. For example, computer time sharing techniques allow for the simultaneous connection of many users to a single remote computer through the use of public or private communications paths. Also, recent innovations permit several computers to be connected into a network using high speed data transmission paths. Such a network can be used to share workloads among several computers or to take advantage of specialized capabilities of particular computers connected to the network. This convergence of computer and communications technologies continues rapidly and is only beginning to influence the major legal framework designed to allocate communications functions.

Telephone and telegram services were first included in federal regulation in 1910 through an amendment to the Interstate Commerce Act [5]. This amendment placed telephone and telegraph services under the jurisdiction of the Interstate Commerce Commission, a body created primarily to regulate transportation services, particularly railroads. In 1934, the Communications Act transferred regulation of communications to the newly created Federal Communications Commission. To a surprising degree, the part of the Communications Act dealing with point-to-point communications consists of language transferred almost verbatim from the Interstate Commerce Act. This language remains central law for point-to-point communications regulation today and results in attempts to alter and rearrange legal concepts essentially designed to control nineteenth century barge and rail traffic so that they cover time shared, message switched, computerized data processing networks.

One apparent motivating factor for regulation of communications carriage services has been the belief that these important services exhibit the economic characteristics of natural monopolies. While regulating monopoly markets is intended to protect the public, it also protects from competition those activities of the communications common carriers that, while incidental to carriage, may not themselves be natural monopolies. The market for computer services and data processing, on the other hand, is currently seen as suitable for competitive, unregulated operation.

Legal, economic, or institutional categories, however, may have only peripheral bearing on technological development. Technological advances in communications and data processing have not "cooperated" with the law. As computer and communications services have developed, they have become dependent on similar technologies, bringing their different market structures into contact. The line separating these services—itself based on technological considerations—has become as much a barrier to understanding as an aid to decision.

When communications regulation first began, its main focus was on regulating signal carriers, and the scope of FCC power over point-to-point communications services may have seemed well defined. Regulation of incidental services was included since many such services were barely perceptible or identifiable from signal carriage. The telegrapher encoded, routed and forwarded the message; the telephone operator provided directory assistance and circuit switching; it was all in a day's work. As computers have been introduced, however, incidental services have come into their own as perceptibly and identifiably separate and distinct from the mere carriage of signals from place to place. The prominence in fact of incidental services is in sharp contrast to their obscurity in the laws and regulations of the FCC. Regulation of incidental services might now cover entire industries.

This article concentrates on the allocation of services which traditionally have been treated as incidental to carriage. Often, these are services with which the user most interacts—the encoding and decoding performed by the telephone instrument or computer terminal, the directory assistance service, and billing are prime examples. Policy for allocating these services affects the nature and rate of the introduction of new services, the rate at which new technologies are introduced to support communications services, the cost of old and new communications services, and, ultimately, the control of communications channels and information flow.

In this article, the allocation system of the FCC serves as an example of a legal and institutional framework straining to adapt to changing perceptions of services and functions. The use of the FCC as an example is purposeful: one cannot discuss the increasing interdependence and technological convergence of computer and communications services without an in-depth examination of the role of the FCC.

Section II treats the FCC's attempt to set criteria for allocating newly perceived and implemented services based on converging computer and communications technology. In 1966, the FCC started an inquiry into the interdependence of computer and communications services and facilities [6]. In the resulting rule making procedure, the Commission set guidelines and categories for determining which services it would regulate and which it would not. The FCC called "common carriage" and "hybrid communications" those services it chose to encompass within its jurisdiction, and labelled as "data processing" and "hybrid data processing" those services it chose to exclude [7].

Significantly, the distinction between these services cannot be based on the technologies they use. For example, a remote access data processing service, which is not to be regulated, might rely on a central computer connected to remote terminals by common carrier network paths. A message switching network, used for quite a different purpose, is to be regulated even though it relies on the same technological configuration. Thus, the FCC guidelines require close examination of the operations performed or the functions served, rather than examination of the technologies involved.

Shifting emphasis from technology to function poses difficulties in the FCC allocation process. When the Communications Act was passed forty years ago, services to be regulated were technologically or institutionally fenced in, and the decision to regulate a service or not could be based primarily on an examination of the technology used or the organization that used it. Computer services were initially supported by distinct technologies and organizations. Hence, computer services were easily excluded from FCC regulatory jurisdiction.

Yet as computer technologies are introduced into communications services, the boundary of FCC jurisdiction becomes increasingly arbitrary and difficult to define. Section III discusses three new or proposed services—separately charged directory assistance, "911" emergency calling with Automatic Location Identification, and packet switched computer networking—that are at the very fringe of FCC regulatory purview. These services are chosen because an analysis of them highlights the difficulties and inconsistencies to be expected as the FCC attempts to implement its allocation policy on a case-by-case basis. At stake are issues of how the FCC procedures for allocating information services and defining its own jurisdiction are to be controlled.

The issues thus raised lead to a discussion, in Section IV, of the development of the notion of communications common carriage. This concept is at the center of FCC regulatory jurisdiction over point-to-point communications services. Determination of the scope of communications common carriage may seem at first glance to be largely a definitional problem. However, attempting to define this concept lends insight into the kinds of services that policy makers outside the FCC thought the Commission should allocate and regulate. Defining communications common carriage also helps to describe which services are to be considered as incidental to carriage—services that, in our observation, demand separate evaluation for their need for regulation.

Whether or not to regulate the functionally identifiable services that are incidental to the carriage component of a communications service is an immediate policy problem. This problem need not be addressed only by the FCC. Congress and the courts, for example, also have roles in making and applying allocation policy. Section V deals with some of the issues, questions, and consequences that follow from a consideration of some of the options available to policy makers in their decisions on whether or not to regulate these services. This section of the article is not addressed to the options available to deal with all communications services; only services hitherto called incidental to carriage are considered.

Finally, throughout this article, allocation decisions are approached by first focusing on the identifiable information functions being served. The decision as to who will serve the functions, and how, can then be made by weighing such factors as where control over the function should be placed, how closely allied or related is the function with other information services, what alternate technologies are available at what prices, and whether political or bureaucratic structures will tolerate creation or change in institutions or organizations to serve the function. We do not suggest that such functional analysis is the final or ultimate policy tool. However, we find such a functional approach useful in providing analytical flexibility in dealing with changing relationships between information functions and the means for serving them.

Allocation decisions need not be dictated by the technological or institutional environment. Rather, technologies and institutions to perform information services can be selected or designed to meet desired policy goals of what functions are to be served, who is to serve them, how they are to be available, and for how much.

II. THE FCC INQUIRY

A. Background

In 1965 a controversy arose over what was data processing and what was communications common carriage, and what was the proper status of each before the Federal Communications Commission. The Bunker-Ramo Corporation had for a time operated a computerized service for quoting securities prices. The information was transmitted through facilities leased from American Telephone and Telegraph and Western Union, but the attendant data processing was controlled by Bunker-Ramo. In 1965, the firm sought to expand the service to enable a stock broker who subscribed to the service to send and execute purchase and sale orders to other subscribing brokers [8]. From a brokerage point of view, it was a logical extension. The decision to buy or sell is based on changes in the price of a stock. Indeed, the stock prices already being transmitted were nothing but general offers to buy or sell. The expansion scheme would have had a computer system accepting purchase orders from stock brokers across the country and matching those orders with sale offers from other brokers.

The Bunker-Ramo situation presented difficulties for the FCC. From Bunker-Ramo's viewpoint, extending the service meant offering additional data processing. To AT&T and Western Union, however, the change meant that Bunker-Ramo was seeking to go into the message switching business. Both utilities refused to lease the necessary communications channels on the grounds that message switching was a business for common carriers only [9]. The issue could not be resolved adequately under the FCC regulations then in force.

The proposed Bunker-Ramo stock system was only one of a number of new services and technologies being offered by various regulated and unregulated firms. These new offerings illuminated the need for consideration of the impact of the convergence of computer and communications technologies on the FCC's regulatory framework. By the mid 60s, computer time sharing techniques had developed so that it was no longer necessary for each computer user to keep his own computer [10]. Organizations that had been too remote or too small to have their data processing done by "in-house" computers could now buy blocks of computer time from distant processors as needed. Each user only needed a computer terminal and a communications path connecting him to the remote computer.

As the computer time market grew and competition developed, the costs of connecting the remote user to the computer—the communications costs—became an increasingly significant part of the cost of providing the entire service. If a vendor of computer time wanted to compete effectively in distant markets, he had to be able to get good communications services at costs comparable to those of local firms in the market. It was clear that the provision of remote access data processing depended heavily on the quality and cost of the facilities and services provided by the common carriers [11].

At the same time, the traditional common carriers, particularly Western Union, were growing interested in this apparently profitable data processing market [12]. Those carriers were already using computers in the technical end of their operations to perform services incidental to carriage, such as circuit and message switching and monitoring network traffic and equipment malfunctions. In fact, many developments in data processing and machine computation had their beginnings in such communications applications [13]. By the time of the FCC inquiry, the carriers were also using computers administratively, for billing, payroll, and the like [14]. The traditional carriers hoped to exploit their technical expertise and dependence on computers to enter new markets.

Finally, as in the case of Bunker-Ramo, organizations other than the traditional common carriers were introducting new service combinations based on the power of computers to engage simultaneously in switching and data processing. Data processing per se was not a concern of the FCC, but when processing data began to include message switching, the FCC's interest was aroused [15]. In order to investigate these developments, in 1965 the Commission instituted an inquiry entitled *In re Regulatory and Policy Problems presented by the Interdependence of Computer and Communication Services and Facilities* [16].

B. The Inquiry

The FCC inquiry was originally directed at a variety of technical, legal, and social issues, ranging from the adequacy of common carrier services for data processors to questions of information privacy [17]. The focus of the inquiry narrowed to the two issues of the greatest concern to the communications and data processing industries and to the FCC. These were, first, the nature and extent of common carrier participation in the data processing market, and second, the extent of FCC regulation of data processing [18]. Of these two, it appears that the FCC sought to avoid the second by stating that the data processing industry did not seem to warrant regulation, implying that the Commission had authority to regulate but was exercising forbearance.

As to the nature and extent of common carrier participation in the data processing market, the following options illustrate the wide range of choices available to the FCC:

(1) to allow carriers to offer data processing without restriction as an unregulated activity;
(2) to allow carriers to offer data processing service as an activity subject to FCC regulation;
(3) to allow carriers to offer data processing services through separately incorporated, unregulated subsidiaries;
(4) to allow carriers to offer data processing services through separately incorporated affiliates which could not sell their services to the common carrier [19].

The carriers argued that their commercial provision of data processing would enhance competition in the data processing market. Also, Western Union asserted that if it could offer commercial data processing services, it could use its own computers more efficiently [20]. Drawbacks to this option were pointed out, however. Other participants in the inquiry noted that the dependence of all data processors on the common carriers would give the carriers a competitive advantage in data processing. The carriers could subsidize low prices for data processing services by drawing on protected monopoly profits from signal carriage. The carriers would also be in a position to discriminate subtly, even undetectably, against competing data processors through such techniques as slow deliveries of equipment and service, assignment of inferior quality equipment and lines, and refusals to provide service because of alleged violations of service rules, called tariffs in regulatory parlance [21].

Given these considerations, the FCC decided that the dangers of unrestricted carrier participation in the data processing market outweighed the benefits, but that the benefits could be preserved by endorsing the fourth option above, known as maximum separation [22]. The Commission issued rules to that end, called the Computer Rules, prohibiting large common carriers subject to FCC jurisdiction from furnishing data processing services to others except through separately incorporated affiliates [23]. These affiliates were not to be represented by the parent in any sale or promotion of data processing and were not even to use the name or symbols of the parent. The parents were not to purchase, lease or obtain in any way data processing services from their subsidiaries [24]. As first stated by the FCC, the Computer Rules seemed to permit the carriers to put only their money into a separate data processing firm. The rules, however, did not address the problems posed by the more subtle discriminatory techniques.

It should be noted that these rules do not apply to the companies of the Bell Telephone System, since by antitrust consent decree, the Bell companies may engage only in communications common carrier activities, with a few minor exceptions [25]. Thus, it would seem that the decision of what services the Bell System may provide is made by a court rather than by the FCC. In reality, however, this procedural arrangement would seem to pressure the FCC to define common carriage carefully so that the services the Bell System can provide are clearly delineated to the court. In other words, it seems that as the FCC defines communications common carriage, so too it decides what services the Bell System may or may not provide.

Some carriers and their data processing affiliates appealed the Computer Rules in *GTE Service Corporation v. Federal Communications Commission* [26]. The U.S. Court of Appeals for the Second Circuit ruled in favor of the data processing affiliates, leaving the FCC to the third option above, less-than-maximum separation. In particular, the court ruled that the parent carrier could obtain data processing services from its affiliate. The ban on use of the parent's name or symbol by the subsidiary was also lifted by the court. The court's decision turned on a finding that "the Commission had no jurisdiction under [the Communications] Act or in the cases construing it, to regulate a separate affiliate's business in the data processing market" [27]. In the court's view, maximum separation forced the carrier's data processing affiliates to operate under restrictions not faced by independent data processing firms.

Thus, the court's decision raised again the second issue facing the FCC inquiry—the issue the Commission appeared to avoid—which was the nature and extent of the FCC's power to regulate the data processing services. The FCC inquiry had concluded that there was no "need to assert regulatory authority over data processing services, whether or not such services employ communications facilities in order to link the terminals of the subscribers to the centralized computers" [28]. The industry, the FCC stated, did not meet the FCC's standards for government intervention and regulation:

> Government intervention and regulation are limited to those areas where there is a natural monopoly, where economies of scale are of such magnitude as to dictate the need for a regulated monopoly, or where other factors are present to require government intervention to protect the public interest because a potential for unfair practices exists [29].

The Commission observed that there were no natural or economic barriers to entry into the data processing market, that the number of firms in the field was increasing, that the industry exhibited effective competition; therefore, the FCC concluded, there was no need for regulation at the time. The FCC intimated that the decision not to regulate data processing was a matter for the Commission's discretion. In the future, the FCC said, if conditions in the data processing industry changed, then the FCC might reexamine its discretionary self-restriction [30].

The *GTE* decision, however, bears the strong implication that the FCC's exercise of discretion was in fact a prudent recognition of its own jurisdictional boundaries. If, as the court ruled, the FCC could not regulate the data processing activity of affiliates of regulated common carriers, then no amount of discretion would seem to permit the Commission to regulate the data processing activity of an unaffiliated, independent firm. Regardless of whether the FCC ought not, will not, should not, or cannot regulate data processing, the Commission still had to draw a line between communications and data processing services in order to know what not to regulate.

The definitions used by the FCC in the Computer Rules focus on the function of the questioned service, rather than on the technology employed. This is clear from comparing the FCC's definitions of message switching and data processing, which may rely on identical technologies. Message switching, the Commission stated, is ". . . the computer-controlled transmission of messages between two or more points, via communications facilities, wherein the content of the messages remains unaltered" [31]. Data processing is:

> . . . the use of a computer for the processing of information as distinguished from circuit or message-switching. "Processing" involves the use of the computer for operations which include, *inter alia*, the functions of storing, retrieving, sorting merging, or calculating data, according to programmed instructions [32, 33].

Basically, if the function was message switching, the FCC was interested, if it was data processing, the FCC was not interested [34]. The FCC insisted that message switching was inextricably bound up with common carriage: ". . . message-switching is, in essence, an operation which in inherent in the process of transmitting, by wire or radio, intelligence or data from point of origin to point of destination" [35]. The Commission seemed to fail to distinguish message switching alone as a function identifiable and separate from signal carriage, like data processing. Thus the FCC forced itself to draw the important line between message switching and data processing.

Yet, despite the hopeful tone of the phrase "processing of information as distinguished from circuit or message-switching," the FCC must have realized that distinguishing the functions was not going to be easy. This was particularly so in cases like the Bunker-Ramo stock service, which combined whatever message switching and data processing was needed to get the word out to customers.

Attempting to relieve these difficulties, the FCC established another category of service and drew yet another line. Services that combine data processing and message switching are *hybrid services*. If message switching plays only a minor part in the combined offering, the FCC will not regulate. If data processing plays only a minor part, then the combined service is deemed communications and subject to regulation. Unregulated combined services are called *hybrid data processing*; regulated combined services are called *hybrid communications* [36].

This set of painfully drawn distinctions is used by the FCC to determine the limits of its discretion or jurisdiction or power. As the Commission stated, "In effect, we shall address ouselves to the facts surrounding a package offering with a view toward determining the primary thrust of the service offered" [37]. This is also called the "primary business test." It has a long history in the law, but was first suggested for FCC use in distinguishing communications and computer services by IBM in 1966 [38].

Under the Computer Rules, the FCC has a number of decisions to make on each new service offering, depending on the nature of the service and whether it is being offered by a regulated or unregulated firm. If an unregulated firm comes before the FCC with a proposed hybrid service, the FCC determines whether the primary thrust is communications or data processing. If it is communications, then the service can be offered only with FCC approval and regulation. If the primary thrust is data processing, then the FCC will not regulate the service. If a regulated common carrier proposes a hybrid service with a primary thrust of communications, then the carrier can offer it under normal FCC rules. If the primary thrust of the hybrid service is data processing, then the carrier can only offer it through a separate entity. The result is that the FCC now regulates some commercial data processing some of the time in some ways.

C. Analysis of the Rules

The FCC allocates the right to carry signals from point to point by wire or radio through procedures established by Congress in the Communications Act of 1934. The Act has its roots in amendments made in 1910 to the Interstate Commerce Act [39]. The Computer Rules promulgated in 1971 were devised to aid the FCC in regulating and allocating services employing computer technology.

The Computer Rules, however, unwittingly or not, widen the FCC's purview over point-to-point communications to include a good deal that is not strictly point-to-point communications. The FCC has always permitted the common carriers to do some things that were not strictly carriage, so long as those things were incidental to carriage. If the data processing is incidental to signal carriage, then the carrier may do it and the FCC may regulate it. But the determination of what is incidental to carriage is not always simple.

The difficulties that follow from altering and rearranging legal concepts essentially designed to control nineteenth century barge and rail traffic in order to cover time shared, message switched, computerized data processing networks are recounted in detail in a later section of this article [40]. We must first examine some of the difficult issues raised by the inclusion of certain modern data processing services within the FCC's purview through the Computer Rules. We look at the enforcement mechanism of the Rules and then at the implications of the primary business test.

1. The Enforcement Mechanism. Having drawn a line between common carriage and data processing, the FCC appears to be relying on the communications common carriers to police that line:

> We anticipate, therefore, that the communications common carrier providing the communications portion of the hybrid service will question any service that appears, in fact, to be a hybrid communications service. This is so because of the tariff provisions of the common carrier that prohibit the resale of service by its customers. The carrier in such circumstances could either call this matter to the Commission's attention or, pursuant to its tariffs, treat it as an attempt for the resale of communications service, which as noted above the tariffs prohibit, and refuse to provide the facility [41].

This assumes a great deal.

If the carriers were not relied on to sound the alarm, the FCC might be able to reach and evaluate new services only with considerable difficulty [42]. As it stands, the FCC appears to be looking to the following scenario. A new service employing common carrier services is offered. The carrier evaluates the new service. If the carrier thinks the service should be regulated, that is, that the primary or only thrust of the service is communications, then the carrier has two choices. It may report the service to the FCC, or the carrier may refuse to sell carriage services on the grounds that the new offeror will be reselling carriage, which is prohibited by the carrier's tariffs. The offeror of the disputed new service, on being rebuffed or reported by the common carrier, can either discontinue his plans or complain to the FCC about carrier unfairness [43].

Or, to avoid confusion, to avoid possible untimely refusal of service by the common carriers, or to avoid finding themselves suddenly and unexpectedly before the FCC at some later date, offerors of new services with any significant communications element can actively seek FCC evaluation and adjudication. This fear of uncertainty may in practical terms mean that the FCC will evaluate any new hybrid service before the service can be offered commercially, a considerable extension of FCC power.

The system also places a great deal of power in the hands of the communications common carriers. FCC jurisdiction is limited to common carriage and hybrid communications. But the carriers themselves provide a vital component in any remote access data processing or hybrid service, since by definition almost all of these services require a point-to-point communications capability. The first decision on whether any particular offering is data processing, hybrid data processing, hybrid communications, or communications common carriage rests with the common carrier providing the signal carriage component of the service.

The Computer Rules of the FCC seem to assume that the regulated carriers will correctly apply the FCC's criteria, in a regular straightforward manner, to important customers who are also potential or actual competitors. No recognition seems to be given to the possibility that some hypothetical, unscrupulous carrier might find advantage in this situation. Such a carrier can force a new hybrid or data processor to report for judgment to the FCC, a forum completely familiar to the carrier and possibly strange to the offeror. Finally, it is left largely to the carrier to determine, in the light of its own interests, when to litigate, when to request waivers of tariff provisions to make way for services otherwise prohibited, and when to exercise carrier discretion and remain silent.

2. The Primary Business Test. Under the Computer Rules, both the carrier and the FCC make make case-by-case examinations of each new hybrid offering in order to determine whether the primary thrust of the service is data processing or communications. This test too raises difficulties.

First, the risk of establishing a hybrid or possibly even a remote-access data processing service is greatly increased because of the danger of attracting the attention of the carriers or the Commission. Until these two have decided, the firm does not know for sure whether the new service will be regulated or not. This uncertainty and the costs of resolving it raise market entry barriers and discourage new enterprises. Second, the burden of preparation for hearings and for making decisions will fall most heavily on the limited budgets of the FCC and the comparatively smaller businesses offering the new services. And third, the primary thrust of a business or offering may change, leaving the offeror continually subject to reexamination by the carriers and the FCC [44].

Also, the kind of black-or-white, sheep-or-goat categorization used by the FCC provides incentives for arbitrary or distorted results. As one scholarly observer of the FCC put it:

> Communication and computer firms will obviously employ the test to their own ends, however those ends are defined. That such a test borders on the arbitrary can be seen in its application. In the eyes of Western Union, flower and data processing service are 'incidental' to communications. Yet the communication industry becomes much less tolerant when the identical test is invoked by the data processing industry to wit, the Bunker-Ramo case.
>
> Thus, the primary business test cuts both ways; it can be employed to rationalize a movement from communications to data processing or from data processing to communications. And if such a test is resurrected as a policy guide, it takes little imagination to visualize a backlog of adjudicatory proceedings in which the adversaries seek the magic number of 49% [45].

Finally, the criteria for determining the primary thrust of a service are not clear and may become less and less clear as the technology develops. Any major differences between hybrid communications and hybrid data processing lie in the functions served; the technology employed will be either very similar or identical. And there are social and economic considerations—that is to say, political considerations—that have hardly been taken into account at all.

These are not conjectural difficulties. They are all too real. The next section examines three new service offerings that demonstrate concretely the uncertainties created by the Computer Rules and the interdependence and interchangeability of computer and communications services and technologies.

III. THREE NEW OR PROPOSED SERVICES, CASE-BY-CASE

A. Separately Charged Directory Assistance

The first telephone system was simple. Alexander Graham Bell was in one room with the first telephone; Mr. Watson was in another room with the second telephone; the two instruments were connected by a pair of wires. The earliest commercial installations were similar, a pair of telephones connecting two places that needed to communicate frequently. The one and only path was there before the subscriber picked up the telephone [46]. Even the introduction of the telephone exchange did not make the system much more complicated. The caller would simply ask for his party by name; the operator was supposed to know the location of each subscriber; and the telephone directory was only a list of subscribers. Who would have thought of the provision of information about the location of subscribers as a function or service identifiably separate from the service of making the switchboard connection?

With the growth of the telephone system and the interconnection of exchanges, the task of the operator became more burdensome. Although directories included numbers by this time, many callers still needed help in making their calls and the exchange operators were tied down looking up numbers. The solution was the organization of a separate "information" service, which would provide users with telephone numbers and leave the exchange operators free to make the connections. As the system continued to grow, the information system grew as well, with massive directories brought up to date daily, thousands of calls being handled hourly, the institution of a national number for directory assistance, and, in general, a great deal of information storage, sorting, and retrieval—that is to say, data processing.

The information service now known as "directory assistance" has traditionally been considered part of the voice carriage network, necessary to the orderly provision of telephone service. Directory assistance is everywhere treated as a service incidental to voice carriage, and the costs of the service have been divided evenly among subscribers. It is operated by the common carrier and protected from the competitive market. Recent developments, however, indicate that directory assistance is being "unbundled" from other voice carriage services and given a rate structure of its own while its classification as incidental to carriage is retained. No one has proposed formally that the telephone company should get out of the directory assistance business; but the very fact that it is being treated in some ways as a separate business raises some interesting questions about incidental services and about the definition of communications common carriage.

The occasion for raising these questions is the newly expressed desire on the part of some telephone companies to charge separately and by the call for directory assistance. Cincinnati Bell already charges 20¢ for each call to directory assistance [47]. The New York Telephone Company (NYT), also part of the Bell System, recently received approval for a similar plan from the New York State Public Service Commission (PSC)[48]. Before the New York plan is implemented, each NYT subscriber will continue to pay approximately 90¢ per month for directory assistance, but as an undifferentiated portion of his monthly fee. Each call to "411" costs 16.5¢ according to NYT, but the company has found that only 10 percent of the subscribers makes some 60 percent of all the "information" calls [49]. Many of these calls are unrelated to the efficient operation of the telephone system. Many calls come from private firms, such as credit agencies, who use the system to check addresses or to find out if a credit applicant has a phone, indicating that he has paid his bills [50].

The plan approved by the PSC assumes that the average customer makes three calls for directory assistance per month [51]. Three calls per month will be allowed without extra charge, but anything over three will be charged at the rate of 10¢ per call [52]. The charge will be the same whether the requested number is in the subscriber's local directory or in another NYT directory [53]. The service charge for directory assistance will be listed separately on the monthly bill [54]. (The subscriber will start the month with a credit for three calls, taking care of "average" users) [55]. The following calls are to be exempt from directory assistance charges: calls from coin telephones; calls placed by handicapped persons who cannot use a telephone directory; calls placed by subscribers of independent telephone companies into NYT territory; interstate calls; and calls for numbers which are, in fact, unpublished [56].

Thus, while directory assistance is officially categorized as an incidental service, both the New York State Public Service Commission and NYT seem to be treating it as a separate commercial data processing offering. The PSC stated that it was allowing NYT to charge for directory assistance on a per call basis in order to distribute the charges for the service more equitably among telephone subscribers [57]. But the New York Commission also concluded:

> In view of the trend of ever-increasing costs, it has become increasingly important to "unbundle" charges for distinct telephone services. In this way, the basic charge for the service can be kept as low as possible and additional services, if required, can be separately purchased by the appropriate subscribers [58].

The history of the "information service" and our experiences using it tell us that the service is incidental to voice carriage, but the PSC decision implies it is not.

NYT has proposed, in effect, to sell directory assistance to customers separately from telephone service. While this characterization may be taken as extreme, it is unquestionably the case that directory assistance is data processing, that it is a service identifiably distinct from the service of carrying voice signals from point to point, and that the telephone company sees some separate commercial possibilities in the service.

If the telephone company sees these possibilities, it is not unlikely that others see them and would be interested in going into the directory assistance business themselves if the phone company did not have a corner on the market. The telephone company wishes to sell information about its subscribers and their standing with the company at 10¢ per inquiry (6¢ if requests are made in writing in bulk, according to the PSC plan) [59]. Should this be a part of their monopoly? If there is to be a separate charge for directory assistance, then why not a separate charge for directories as well? If the telephone company can distinguish the cost, functions, and benefits of directory assistance with such precision, and if the commercial possibilities are so attractive, what then is incidental about the service? Granted, long distance information probably is a natural monopoly, but is it necessary that the telephone company operate it? And on the local level, equipment designed for Automatic Location Identification (explained below) [60] could make an open competitive market for directory assistance possible. Local directory assistance has many different uses and different firms might offer it, once having repaid the phone company for the cost of gathering and updating the information.

Certainly, this sort of speculation will be viewed with alarm by the common carriers and their regulators. But the point here is the value of stepping back from the technology to examine the function that the technology is performing. Just because a service was once incidental to common carriage is not in itself sufficient reason to continue treating it as such. The service of directory assistance, which on first appearance seems virtually an integral part of communications common carriage, turns out on closer examination to be something of an independent opportunity for some telephone companies. And if directory assistance is not so incidental anymore, the case for treating other, more complex data processing services as incidental to carriage may prove to be difficult to make, particularly if we continue to examine functions rather than technologies. The following example indicates some of the nontechnological problems that can arise with the introduction of new technology.

B. Automatic Location of Callers Summoning Emergency Aid

In 1968, at the height of public concern about "crime in the streets," AT&T announced that it was setting aside the telephone number "911" nationwide for use as a universal number to call for emergency aid [61]. General Telephone and Electronics, the second largest phone system operator in the country, also committed itself to the idea [62]. Without question, it is difficult to find the right source of help in an emergency. Even in the largest cities it is normal for different agencies to provide fire, police, and ambulance services; and most suburban areas present surprisingly tangled emergency service arrangements with incorporated and unincorporated areas; fire protection districts predating suburban development; services provided by contract between towns; ambulances provided by hospitals, police, fire departments, and private services; highways patrolled by state police, roads patrolled by sheriff's deputies, and streets patrolled by local police. The ordinary citizen in an emergency is likely to be excited, confused, and hardly ready to sort out complicated jurisdictions or to analyze just what sort of help is needed to deal with the emergency at hand. A combination of technology and organization to support a single emergency phone number was suggested to eliminate some of the confusion.

Response to the "911" idea was immediate. By May 15, 1973, there were 239 "911" systems in operation [63], paid for by the communities and the telephone subscribers using them [64]. As recently as November 1973, Dr. Clay T. Whitehead, Director of the White House Office of Telecommunications Policy, reiterated the federal government's interest and encouragement of the adoption of a nationwide emergency telephone number system [65].

The "911" idea is not at issue, but the problems raised by actual implementation of the service illustrate the difficulties the FCC faces as computers become more and more important in communications. Present "911" systems are simple. Dialing "911" connects the caller to a special emergency services operator, who determines the nature and location of the problem and passes the information to the appropriate agency or agencies [66]. Two problems with the simple system are the presence of a middleman between the emergency and the assistance and the frequent difficulty in getting an excited, frightened caller to give a precise location. While these difficulties are not insuperable, they place a practical limit on the area and population each "911" system can serve, and they put heavy burdens on the ingenuity and capability of the emergency services operator.

More advanced technologies are available for "911" use. These will automatically identify the location from where the "911" call is being made and will automatically route the call to a dispatcher covering the geographic area or political subdivision in which the caller is located [67]. Under these systems even a small child dialing "911" would unknowingly provide much of the information the dispatcher needs, and even a severely disabled person could summon help.

These advanced technologies support what are called Automatic Location Identification (ALI) systems. At present these systems are available in two forms: centralized and on-line. The centralized system incorporates telephone company equipment already used for automatic billing of toll and long distance calls. The on-line system would require installation of new equipment at every telephone instrument. In either case, a "911" system with ALI and automatic routing would include a significant amount of data processing.

In the centralized system, the calling number is identified by the billing equipment, and transmitted to a computer data base which gives the actual location of the telephone. The call is then transferred to the appropriate emergency dispatcher [68]. With the on-line system, each telephone instrument is fitted with a programmed integrated circuit—in effect a tiny computer—which automatically transmits the location when the "911" call is answered, permitting an automatic switching facility to route the call [69].

Especially in the case of the on-line system, it is not technically necessary that the ALI system be operated entirely by the telephone company [70], but just who might get to offer this data processing service raises several issues under the FCC Computer Rules. General Telephone and Electronics recognized the problem and requested an FCC staff ruling on whether or not, and under what conditions, common carriers could offer ALI as part of a "911" service [71]. The FCC's Common Carrier Bureau concluded that the data processing involved is "for the purpose of providing an essential informational service to '911' subscribers that is incidental to the carrier's basic telephone service [and] . . . [is] permitted [for a common carrier] under the Commission's Computer Rules . . ." [72].

This initial observation of the Common Carrier Bureau is not without difficulties, as a closer comparison of the different ALI schemes reveals. First is the question of intertwining this new service with the telephone companies' voice carriage service. Centralized ALI, using automatic billing equipment, would significantly tie ALI operations to the equipment and procedures of long distance service. On-line ALI, with each subscriber having his own electronic "address label," requires the telephone company to provide wire interconnections only. The "foreign" attachments needed to implement the on-line scheme could be provided by a non–telephone company entity—a proposal made possible by the FCC decision in the *Carterfone* case [73], allowing non–telephone company devices to be connected with the telephone network, subject to network technical requirements.

The two ALI schemes also differ in their economics. The centralized system seems likely to have generally lower costs, chiefly because of ease of maintenance [74]. This is true, however, only in large communities or where the automatic billing equipment is already installed. In smaller communities that do not have the automated billing setup, the on-line system would be cheaper to start with, although as the system grew, the costs of equipment and maintenance would pass those of the centralized system [75]. The economics of on-line ALI are not frozen forever, however; production of integrated circuitry is quite new, with volatile economics. This is illustrated by the rapidly falling prices for pocket electronic calculators, which use similar technology. There are performance differences between the two schemes, as well: centralized ALI using present billing equipment cannot distinguish among different subscribers on a multiple party line [76]; on-line ALI, of course, is designed to do just that.

None of these points is decisive, however. The aim is to provide a single number and a single system for summoning emergency aid. The job of connecting the caller with the emergency service center will be handled by the telephone company under any system. It might therefore seem natural to allow the phone companies to provide the necessary data processing as well. It might also seem natural and sensible to take advantage of previous technoligical developments, to use the phone company's billing equipment and to do the job with centralized ALI. But at least one problem remains. A survey of municipal public safety officials on "911" implementation conducted by the Franklin Institute Research Laboratories found that "[o]ne of the major objections to a single emergency number involved the conflicts or number of problems that arise over jurisdiction" [77]. Some of the survey respondents even expressed fear of "empire builders."

Turning over front line control of their fire and police services to a central dispatcher not directly answerable to them might trouble many municipal governments. Since most telephone companies are regulated only on the state and federal levels, local officials foresee some of their control slipping away if the "911"-ALI system is too closely tied to the telephone company. Thus, in the eyes of local authorities, the greater technical, economic, and administrative separation possible under the on-line ALI system might weigh heavily in its favor.

On another political level, allowing common carriers to provide ALI as an incidental service would greatly expand the restricted role these carriers are supposed to play in the data processing market. Present data processing by such carriers already includes weather and time services as well as directory assistance. All of these, and ALI as well, are qualitatively different from the functionally incidental data processing (exemplified by billing or studies of telephone traffic) necessary for the operation of the telephone network.

A final question about just how "incidental" is automatic location identification concerns other possible uses of the data processing capacity of a tiny computer at each phone. A centralized telephone number identification scheme is already in use by the phone company for toll calls and billing. On-line ALI, however, is far better suited to other possible applications such as credit verification, facsimile transmission, computer access, directory assistance, and any of the myriad of other possibilities that might be imagined as part of an on-line system. Should these services be offered solely by the phone company, solely by competitive service vendors, or by joint enterprises between the phone company and the vendors?

The FCC's original ruling then leads to precisely the same sort of dilemma that prompted the FCC to begin its data processing inquiry in the first place. This results directly from the tradition of assuming that the allocation of signal carriage automatically allocates many other communications services. Both on-line and centralized technologies provide automatic location identification, but the differences between them demonstrate concretely the different social policy results that can flow from an apparently technical decision.

The on-line technology separates ALI from other telephone company services. This means that ALI can be discussed in terms of the costs for local jurisdictional control, how much control over information functions should be vested in regulatory agencies or regulated monopolies, and, indeed, what policy toward monopolies should be. The technology can be chosen after the criteria are set, rather than setting the criteria in terms of the technology.

C. Packet Communications

On close examination, neither directory assistance nor ALI is as closely tied to the provision of telephone service as it might appear at first glance. These services have come to be considered incidental to carriage for at least two reasons. First, they have always been considered incidental. Second, "411" and "911"-ALI seemingly amount to "accessories" to the larger telephone system, incidentally provided in much the same way that automobile manufacturers incidentally sell air conditioners, radios, and vinyl upholstery. In the auto business, however, there is a full free market for accessories, while prior to the *Carterfone* decision, the telephone companies were able to prohibit even so mild a variation as metal and plastic discs used as dial attachments on telephones if the attachments were not provided by the company [78].

Analyzing the functions carried out rather than the technology employed demonstrated that there was considerable doubt that the telephone company must control "411" and ALI in order to provide communications common carriage. Both services involve considerable data processing, and defining them as incidental extends the influence of the common carriers and their regulators in the data processing business. The FCC has attempted to draw some lines that will keep it from encroaching on data processing, outlining the categories of communications common carriage, data processing, hybrid communications and hybrid data processing.

It is not clear that technological and commercial developments will stand still long enough to be categorized by the FCC. The categories outlined are lacking in precision, and "411" and Automatic Location Identification could probably be included in either of the FCC's schemes for data processing or common carriage without any untoward wrenching of definitions. But regardless of category, directory assistance and ALI arguably are distinctly peripheral enterprises. The Commission, however, has recently applied its analytical powers to a proposed service that directly challenges the very definition of communications common carriage.

Existing organizations, even the most intensely technological, often have difficulty in absorbing fundamental technological change. One hears stories that after World War I, the armed services of the world earnestly debated whether the tank was a mechanical horse or a moving trench; that is, whether the cavalry or the infantry would be in charge of tanks. It is also said that after World War II, the debate was whether the rocket was an airplane or an artillery shell. For many years, transistors were used only as smaller, cooler vacuum tubes. But the realization came that while transistors had a superficial resemblance to tubes and could do many of the things that tubes could also do, transistors in fact marked the beginning of a new kind of electronics. The application to provide commercial service from Packet Communications, Inc. [79], recently ruled on by the FCC [80], raises questions of the same order.

Packet Communications, Inc. (PCI or Packet) is offering a new kind of commercial service called packet switched computer communications. Packet's application posed considerable problems for the FCC's categories, as packet communications is a protean concept, changing shape each time it is viewed. It is entirely data processing and entirely communications, it is all hybrid and yet the FCC has decided that PCI is nothing but a common carrier and may offer no data processing at all [81].

No amount of explanation will resolve the paradoxes presented by the Packet application and decision, but some rough, simplified technical background may help explain the FCC's difficulties [82]. Packet switching is a technique, to be employed by PCI, to facilitate the operation of a network of computers. The network consists of computers located at geographically distributed points, or nodes, connected by communications paths. In PCI's proposed network, there will be at least two possible paths between any two nodes.

For purposes of explanation, we can assume that there are to be two computers at each node of PCI's network. One, a "home" computer, is the computer operated by the subscriber to the network; it might be a general purpose or specialized computer, and it can be different from any other computer in the network. The second or "auxiliary" computer is operated by Packet; this computer is similar to the auxiliary computer at every other node in the network. The communications paths for carrying signals from node to node are, in Packet's network, to be provided by the established communications common carriers, not PCI.

It is perhaps easiest to understand the operation of PCI's packet switched network by following a simplified version of what happens to a message as it travels from a home computer at one node to a home computer at another node. The message between the computers is in "bits," or electrical representations of the binary digits "0" and "1." The originating home computer begins by delivering the message to the auxiliary computer at its node along with instructions as to the final destination. Packet's auxiliary computer then goes to work. It divides the message into packets of no more than approximately 1,000 bits each. Each packet is then set into the proper format with, among other things, a label giving that packet's number in the order of packets and the final destination.

As the auxiliary computer prepares to give a packet to the carrier for shipping to the next node down the line, the computer checks its memory for network traffic information. If the next node is busy or out of order, the sending auxiliary computer can reroute the packet. As the packet leaves the originating node, the auxiliary computer generates error checking data, which is included at the tail end of the packet as it leaves. The sending computer stores a copy of the packet just in case something goes wrong.

At the next node in the trip to the final destination, the auxiliary computer checks the packet for errors, using the error check data sent along with the packet. If no errors are found, the computer sends an acknowledgment to the node from which the packet was received and forwards the packet to the next node on the way to its ultimate destination. If errors are detected, the auxiliary computer throws away the packet. When no acknowledgment is received, the auxiliary computer that sent the discarded packet makes another copy of the packet and sends it again.

After a series of these hop-by-hop transfers, the packet arrives at the auxiliary computer at the final destination. At each node along the way, the packet has been checked for errors and possibly rerouted, depending on network traffic. In the meantime, all the other packets in the original message were making a similar trip. The auxiliary computer does not have to wait for the first packet to make it even to the first intermediate node before sending the next packet toward the final destination along some other route. Since each packet is sent separately and routed individually, the packets may arrive at the auxiliary computer at the final destination out of their original order. The auxiliary computer at the final node puts the packets back into their proper order, deletes all the error check, routing, and format data, and then delivers the message to the home computer at the final destination.

PCI has been called a "value-added" service, since it takes the service offered by signal carriers, adds to that service the operations of the auxiliary computers at each node of the network, and sells the combined offering. It can be seen that PCI processes the data given to it, then turns that processed data over to the carriers for carriage. But the FCC has ruled that Packet is a common carrier and may offer no data processing.

A packet switched network offers many possible uses not available with conventional remote data processing [83]. In a packet switched network, the home computers are owned by the users of the network. With a national network in operation, a national firm could use it to connect regional computers together. For all practical purposes, the firm would then have a single large computer, since a packet switched network like the one PCI proposes will take less than one-half second to get a message from any one node to any other. Or, one network user could trade his computer's large capacity memory with the exceptionally fast processing capability of another user's computer. There could be many such private arrangements between users made possible by the interconnections provided by the network.

Another possible use of a packet switching network would be the marketing of specialized computer services to other users of the network. A firm might, for instance, develop an unusually effective computer program for handling payrolls and pension plans. Since unauthorized computer program use is difficult to police, the program developer naturally would be reluctant to let such valuable software out of his control. A packet switched network can provide other firms with easy access to the desired software while allowing the developer to protect his secret at his home computer. If the development of PCI's service proceeds as expected, the result may well be a nationwide community of computer users.

One particular feature of PCI's network illuminates the extent to which it is offering a data processing service despite the FCC's classification. PCI's computers will constantly check the packets for errors. If any are found, the entire packet is given to the carriers to ship again, and the process is repeated until the data are received error free.

The FCC was introduced to these issues by an application to provide commercial service filed by PCI. The firm seems to have concluded that the Commission's Computer Rules might or might not require FCC approval of the service [84]. The application was an easy way to resolve the uncertainty. As a data processor or hybrid data processor, PCI would be unregulated. As a communications carrier or hybrid communications service, Packet would be subject to FCC regulation.

In response to the Packet application, Western Union [85] and several other regulated carriers, including Microwave Communications, Inc. (MCI) [86], and Data Transmission Company (Datran) [87] argued to the FCC that PCI should be regulated as a common carrier. Their position was that Packet would in effect be leasing lines from carriers and reselling them. Under the FCC's decision in *Mackay Radio and Telegraph Company* in 1938 [88], leasing and reselling lines is a service limited to common carriers. In *Mackay*, however, the parties had stipulated that they were common carriers subject to FCC jurisdiction, while PCI originally made no such stipulation [89].

As a further reply to the Packet application, MCI asked that the FCC treat the application as a request for a rule making procedure. MCI was asking that the FCC establish rules covering services which, like PCI, lease conventional signal channels and add computer services to those leased channels [90]. Although MCI leases some channels from other carriers, it also operates its own channels. Therefore, it was already in business as a regulated common carrier at the time of Packet's application. Owning channels naturally represents a much greater capital investment than leasing them.

Packet responded by stating that it was willing to be classified as a common carrier [91], emphasizing that it had made the application originally to be so classified in order to avoid any lengthy proceedings and to get on with its business:

> PCI believes that the unique aspects of the service it intends to offer should be fully and formally reported to the Commission and, further, the advantages of such services should be brought to the public as quickly as possible. The most expeditious manner in which to accomplish this result is, PCI believes, to file the instant information in the form of a Section 214 application . . . [92].

FCC rule making procedures in such a complicated case could take years, leaving the field open to other common carriers who already have channels and FCC approval of their own. In the meantime, AT&T requested that the FCC waive some of the restrictions in its tariff concerning hybrid service vendors [93]. Packet would be the clear beneficiary of the waiver.

The FCC chose to make a decision before making more rules, and on November 14, 1973, announced that it had "concluded that the present and future public interest, convenience and necessity will be served by a grant of PCI's application" [94]. The Commission observed that PCI was offering a new and improved means of data transmission to users and that it could find no justification for delay. Packet's service was classified as common carriage under § 3(h) of the Communications Act [95], subject to regulation by the FCC under Title II of the Act.

The FCC decision seems warranted, at least at first glance. Packet proposes to accept messages and transmit them to a distant point, essentially unchanged in content. This could be considered as offering a signal carriage service for hire and seems to fit the description of a communications common carrier. Certainly, the FCC's decision solved some immediate problems, but it may have created some future problems as well. If we look closely at the functions performed by Packet, there is room for substantial doubt that PCI is a common carrier.

In terms of its components, however, Packet's service is relying on established common carriers to perform the actual carriage of signals between the nodes of the network. Packet is to provide only the computational power needed to complete the packet switching offering. PCI is not proposing to perform any signal transmission or carriage [96]. And the distinction between PCI's computational element and the signal carriage provided by the carrier is clear.

According to the FCC's Computer Rules, a service which includes, inter alia, sorting, merging, storing, retrieving, and calculating data is a data processing service [97]. These are precisely the services Packet adds to the operation of the carrier network. Packet takes the simple carrier service and adds its specialized data processing service. So, it might be said, do telephone answering services or national news wires, neither of which are subject to the FCC.

In the following section we take the FCC's characterization of PCI as a communications common carrier and test Packet's service by the traditional legal definitions of common carriage.

IV. COMMUNICATIONS COMMON CARRIAGE AND PACKET SWITCHING

A. The Common Law: Divided Responsibility

The FCC has characterized Packet Communications, Inc., as a communications common carrier. Packet proposed to offer a very new service. Communications common carriage, however, grows out of common carriage, a very old concept.

Basically, a common carrier is one who must carry for everybody who can pay the freight. Government regulation of common carriers traditionally consists of making sure that the carriers do in fact carry, and that they are in fact available to everybody on equitable terms. This idea lies behind government regulation of transportation as well as communications. But most of the legal concepts governing this regulation were developed with only transportation in mind simply because there was hardly a perceptible communications industry while the law was developing. For this reason, the FCC has often found itself with a regulatory structure that pretends that communications channels are railroad tracks.

The United States Supreme Court long ago defined a common carrier as "one who undertakes for hire to transport the goods of those who may choose to employ him, from place to place" [98]. A common carrier has special common law obligations to his customers; he must insure the people or goods he carries: "A common carrier . . . has so complete control, and the consequences of negligence on his part may be so serious, that he is justly held to a very high degree of . . . safety" [99].

A common carrier cannot easily escape this common law responsibility. For example, it has been held that as a matter of common law, one need not own the means of transportation to be a common carrier [100]. If a warehouseman contracts for the transportation of another's goods and has complete control over those goods from the time they are delivered to his warehouse, then he is a common carrier even though somebody else is doing the carrying. The warehouseman as a "transportation broker" cannot divide his responsibility for the safe carriage of the goods from the responsibility of the person he hires to do the work.

Applying these transportation concepts to communications services has always been troubling. In 1893, in *Primrose v. Western Union* [101], the U.S. Supreme Court ruled that the telegraph company was not a common carrier:

> Telegraph companies resemble railroad companies and other common carriers, in that they are instruments of commerce. . . . But they are not common carriers; their duties are different, and are performed in different ways and they are not subject to the same liabilities [102].

The question of liability most concerned the Court as it provided two reasons for distinguishing telegraph companies from railroad companies in the *Primrose* decision. First, a telegraph message was particularly prone to mistakes in carriage, no matter what care the company took with the then available technology. Second, the Court reasoned, a message, unlike goods, cannot be embezzled and has value peculiar to the user [103].

Today communications common carriers still do not guarantee completely accurate signal carriage without additional fee. This fact provided the incentive for Packet to provide an error checking capability. Further, in its *Carterfone* decision, the FCC implied that analogy between a communications service offeror and the warehouseman may not stand either. In *Carterfone* the FCC ruled that communications equipment provided by unregulated firms could be connected with the public communications network, subject to technical restraints [104]. This opened the way for private firms to offer various communications services and technologies using the public network, in competition with the regulated carriers.

The *Carterfone* decision also might be interpreted as stating a policy that allows a service offeror who merely interconnects communications paths to divide his responsibility for proper service from the carrier's responsibility for safe carriage. Indeed, Packet's proposed tariff stipulation to that effect has remained unchallenged: Packet "is not liable for any act or omission of any other company or companies furnishing portions of the service" [105]. Thus, concepts of liability developed in transportation contexts may lead to anomalous results when applied to new, developing communications services, like Packet.

On the one hand the FCC has named PCI as a common carrier because it holds itself out as providing the service of carrying a subscriber's data from one place to another. In analogy to the warehouseman's case, it is immaterial that Packet buys communications channels from others and resells them. On the other hand, PCI has not assumed any of the obligations imposed on the carriers with which it contracts, a stance that seems to be permitted under the *Carterfone* decision.

B. Statutory Creation of Communications Common Carriage

Adding communications services to common carriage created a new class of entities: communications common carriers. The history of this development also provides insight into whether or not Packet fits into the class.

The telegraph and telephone companies were first officially included as federally regulated common carriers in 1910 through an amendment to the Mann-Elkins Act [106], which itself amended the Interstate Commerce Act [107]. The amendment carried despite the uneasiness some members of Congress felt in fitting a communications firm into a transportation concept. Both Senator Elkins and Representative Mann spoke in opposition to the amendment:

> They [the telephone and telegraph companies] are not like railroads, and the laws and regulations that would apply to one would not apply to the other [108].
>
> This bill is drawn for the purpose of transportation of passengers and property. It cannot be fitted to cover the transportation of electric fluids . . . [109].
>
> The provisions of the law under which we authorize the Interstate Commerce Commission to regulate charges expressly provides that we authorize them to regulate charges for the transportation of passengers or property. Now, how ridiculous is it to stick into the amendment something which has nothing to do with either passengers or property [110].

Nonetheless, the Interstate Commerce Commission was to regulate carriage of passengers, property, *and* "electric fluids" for the next quarter-century.

When radio communications developed, it was first placed under the jurisdiction of the Department of Commerce [111] and later under the regulation of the Federal Radio Commission [112]. But with the New Deal reorganizations, all regulation of communications—broadcasting and point-to-point common carriage—was placed under the newly created FCC by the Communications Act of 1934. The Act, however, does not provide any definition of common carriage to apply to Packet beyond that derived from prior legislation or common law: " 'Common carrier' or 'carrier' means any person engaged as a common carrier for hire, in interstate or foreign communication by wire or radio, or in interstate or foreign transmission of energy . . ." [113].

Nor does the legislative history of the Act provide further insight into how Congress intended to modify the existing law about who was to be included in the class of common carriers: "It is to be noted that the definition [of common carrier] does not include any person not a common carrier in the ordinary sense of the term . . ." [114].

Though Congress provided no definition of the class of communications common carriers it had created, it seems clear that, in 1934, communications common carriage was thought to include the total service of a communications system like that of Bell and Western Union [115]. It also seems clear that signal carriage technology of 1934 would not permit many operations incidental to carriage to be identified or perceived as functionally identifiable services, separate from the service of carrying signals. In 1974, however, computer technology does permit service offerors, like PCI, to add functionally identifiable services to the signal carriage offered by traditional common carriers.

The task of defining who are communications common carriers seems to have been left to the FCC. An example of an attempt to provide a benchmark definition came in the Commission's 1958 decision in *Frontier Broadcasting Co. v. Collier* [116]:

> Fundamental to the concept of a communications common carrier is that such a carrier holds itself out or makes a public offering to provide facilities by wire or radio whereby all members of the public who choose to employ such facilities and to compensate the carrier therefor may communicate or transmit intelligence[4] of their own design and choosing between the points on the system of that carrier and other carriers connecting with it. In other words, the carrier provides the means or ways of communication for the transmission of such intelligence as the subscriber may choose to have transmitted. The choice of the specific intelligence to be transmitted is, therefore, the sole responsibility or prerogative of the subscriber and not the carrier.
>
> [4] The term "intelligence" is used herein to denote writing, signs, signals, pictures and sounds unless otherwise indicated specifically [117].

It should be noted that this definition closely follows the transportation model—"one who undertakes for hire to transport the goods of those who may choose to employ him, from place to place" [118]—with "transmit" replacing "transport" [119] and "intelligence" replacing "property."

Applied to Packet's offering, however, this definition can lead to divergent results. The problem comes with the term "intelligence." If we equate intelligence with information or data, then Packet's service is the service of a conventional common carrier. The subscriber chooses the data to be sent, chooses the final destination, and instructs Packet to make sure the information is carried. Neither the sender nor receiver need be aware of the manner in which Packet carries out its task: to them the operation is as "transparent" as that of the telephone company or the post office.

On the other hand, there are numerous phrases in the definition given in *Frontier* that would seem to exclude PCI from the class of common carriers. The subscriber to a common carriage service chooses and designs the signs and signals to be transmitted; this is stated or implied four times in the brief definition of common carrier above. Yet, through its error checking, routing, and packet arranging functions, Packet is modifying the information supplied by subscribers. It is Packet that chooses and designs the specific signals to be carried over the network. The writing, signs, or signals carried are Packet's and might have little or no meaning or value by themselves to either the sending or receiving subscriber.

More significantly, according to the *Frontier* definition, a carrier provides facilities by wire or radio. But PCI does not provide such facilities or perform the carriage in any other way. The carriage of signals and the responsibility for carrying them rest with the established carriers from whom PCI buys service. And those carriers, whose services are functionally identifiable and separable from the computation component of PCI's service, are already regulated by the FCC.

C. The Act of 1934: The "Including" Clause

Although the Communications Act of 1934 gives no definition of communications common carriage, the Act does provide a structure to be employed by the FCC in allocating and regulating services. Since Packet's application to provide services has been approved by the Commission, a closer look at the underlying statute may reveal some of the reasons why the FCC defined Packet as a common carrier.

The Communications Act is divided into six major sections or titles. Title I sets up and states the purpose of the FCC and provides definitions. It and Title II apply to point-to-point communications by wire or radio. Section 1 of Title I, in stating the purposes of the Act, seems to give the FCC an extremely broad mandate:

> For the purpose of regulating interstate and foreign commerce in communication by wire or radio . . . there is created . . . the "Federal Communicatons Commission" . . . which shall execute and enforce the provisions of this Act [120].

The Courts have generally construed this mandate as giving the FCC a very broad jurisdiction [121], reflecting other sections of the Act, § 2 (a) in particular:

> The provisions of this Act shall apply to *all interstate and foreign communication by wire or radio and all interstate or foreign transmission of energy by radio, which originates and/or is received within the United States and to all persons engaged in such communication* . . . [122]. (emphasis added).

Yet the broadness of Title I is in sharp contrast to the narrowness of Title II, which, if read alone, would lead one to believe that in the point-to-point communications context, the Act deals almost exclusively with the telephone and telegraph companies and other common carriers as described in §3(h) [123], i.e. "any person engaged as a common carrier."

(Title III deals with radio communications both point-to-point and broadcast. Title IV is procedural and administrative. Title V gives penal provisions. Title VI is a catch-all giving the effective date of the Act, necessary repeals and modifications of other legislation, and the President's war power over communications.)

Title II applies to common carriers; Packet has been classed as a common carrier, yet nowhere in the Act is "common carrier" defined. The Act does, however, include some description of the activities of people engaged in communications, as in §§3(a)-(b) [124]:

> (a) "Wire communication" or "communication by wire" means the transmission of writing, signs, signals, pictures, and sounds of all kinds by aid of wire, cable, or other like connection, between the points of origin and reception of such transmission, *including all instrumentalities, facilities, apparatus, and services (among other things, the receipt, forwarding, and delivery of communications) incidental to such transmission.*
>
> (b) "Radio communication" or "communication by radio" means the transmission by radio of writing, signs, signals, pictures and sounds of all kinds, *including all instrumentalities, facilities, apparatus, and services (among other things, the receipt, forwarding, and delivery of communications) incidental to such transmission.* (emphasis added).

The question here is the inclusiveness of the "including" clause. Since "transmission" can be equated with "carriage," the interpretation of these sections leads to greatly differing views as to the meaning of communications common carriage, to the accuracy of calling Packet a common carrier, and to the scope of FCC power.

At one extreme, these sections provide a very narrow scope for FCC regulation and a conclusion that Packet is not a common carrier. We start with the assumption that in the point-to-point context, §§3(a)-(b) define only what a common carrier in communication does for hire. If one first "qualifies" for regulation by carrying signals from place to place, then the "including" clause insures that all activities incidental to that main purpose are also regulated. Thus, switching, routing, and forwarding services would be regulated if offered by a carrier.

At this extreme, however, the converse would not apply. If one does not first qualify as a common carrier by carrying or transmitting signals, then performing services covered only in the "including" clause would not make one into a common carrier. Thus, answering services, which receive, forward, and deliver communications by wire, are not covered by the FCC because they do not transmit; while the receipt, forward, and delivery services of Western Union are covered because Western Union transmits (carries) signals as well. Under this interpretation, Packet would be excluded from FCC jurisdiction because it actually performs no carriage.

At the other extreme, however, these sections allow both a very broad scope for the FCC and a conclusion that Packet is a communications common carrier. This results by starting with the assumption that anyone who performs any of the functions listed in § § 3 (a)-(b)—signal carriage, receipt, forwarding, or delivery—would be communicating for hire and would therefore be a communications common carrier. This would give the FCC jurisdiction over services which perform no actual signal carriage. Clearly Packet would fall in this class, and so might answering services, stock tickers, press wire services, and any eventual electronic funds transfer system.

D. Section 214: Lines and Channels

It seems quite plausible to suggest that in order to be a common carrier, one must be common, or open to the public generally, and one must also be a carrier. Since transmission (carriage) over lines is the business of communications carriers, a great deal hinges on what constitutes transmission over lines. Section 214 of the Communications Act deals with extending, constructing, acquiring, operating, or engaging in transmission over lines. Thus, the section may help define what carriage or transmission function one must perform in order to be a communications common carrier and may lend insight into whether or not Packet should be classified as such a carrier.

Section 214 defines a line as: ". . . any channel of communication established by the use of appropriate equipment, other than a channel of communication established by the interconnection of two or more existing channels . . ." [125]. A common carrier, then, operates channels of communications. The question of interconnection is taken up below [126].

Depending on the definition of "channel," Packet may or may not be operating a channel of communications. At least two different definitions lead to two different interpretations of the range of FCC power in dealing with Packet. The more scientific of the two defines "channel of communications" as a means for getting a certain quantity of signals at a certain strength and quality over a certain distance [127]. This definition could cover the air between my mouth and your ear, two tin cans and the piece of waxed string that connects them, or it could cover two towers connected by microwave radio transmitters. The common element is that in technical terms, the idea of a channel is closely allied with the actual movement of signals over a distance. The definition would not seem to include Packet's operations because Packet neither owns nor operates any channels, but relies on others to move signals over the distances between the points in PCI's network.

A broader definition of channel of communication has been provided by Kelley Griffith of the FCC's Common Carrier Bureau: any path using wire or radio over which information flows [128]. This definition makes little functional distinction between the actual movement of signals over a distance (the carriage) and the other services incidental to carriage. If this is what Section 214 means, then offerings that include transmission of information for hire by way of wire or radio—Packet, the Associated Press, Able-1 Telephone Answering Service, Crimson Travel Service—would seem to be subject to FCC power.

It should be observed that all services covered by Title II are not covered by Section 214. The most common example would be local telephone companies: they engage in interstate or foreign communications by wire solely through physical facilities supplied by other carriers. In other words, these carriers are not themselves responsible for carrying signals across state borders. Rather, they connect with other carriers who do the actual interstate carrying, and thus are called "connecting carriers" [129]. A connecting carrier is subject to §§201–205 of the Communications Act. Yet, even though such a service offeror is not subject to the requirements of Section 214, he still must be a carrier before he can be classified as a connecting carrier [130]. Again, one is left with Section 214 to infer what a communications common carrier does and whether Packet is a carrier.

While connecting carriers are not subject to full FCC regulation, people who interconnect channels obviously play an important part in communications networks. For a network to exist, messages or circuits must be switched—i.e., carrier channels must be connected with one another. Computers are increasingly employed to perform these important switching and interconnecting functions. However, a new channel created by interconnecting two or more existing channels is not subject to Section 214. The implication is that while extending, constructing, operating, or engaging in transmission over channels requires FCC approval, merely interconnecting two or more channels does not.

How clearly the excluded function of interconnecting channels is to be distinguished from the function of carrying signals has an impact on how common carriage is defined, and thus on the scope of FCC regulation and the proper status of PCI's service. Again, there are conflicting indicators.

In a recent U.S. Court of Appeals case [131], the FCC argued and won a decision holding that the construction of carrier switching equipment was exempt from the requirements given in Section 214 because of the clause excluding interconnection from the definition of "line." The court, *per curiam*, noted that the "Act's legislative history indicates that this provision was inserted to prevent the FCC from asserting jurisdiction over switching equipment while permitting it to continue regulation of carrier equipment" [132]. This statement suggests that identifiable services, which may be incidental to carriage but which do not themselves perform carriage, are to be excluded from the definition of carriage implied in Section 214. This would leave Packet free from regulation.

In another context, however, the FCC has asserted jurisdiction over a switching service in which no carriage at all is included. According to the Computer Rules, all pure message switching services are to be regulated. But there are at least two kinds of pure message switching services which the FCC did not distinguish. In one, the service offeror himself might lease communications paths between users and a switching center and sell the entire network service as a package, point-to-point communications paths and all. Or, in the other, the service offeror would operate only the switching center and the subscribers would provide their own communications paths to and from the center, presumably by buying time and capacity from a common carrier.

Regardless of this distinction based on who provides the communications paths, the FCC asserted authority over the ARX message switching service operated by International Telephone and Telegraph (ITT) [133]. The FCC so ruled even though ITT is not a domestic common carrier and even though the ARX domestic service involves only a time shared message switching computer to which the subscribers lease their own lines [134]. If ARX, with no carriage at all, falls under FCC jurisdiction, then Packet would too.

E. Some Congressional History

An examination of the legislative history of telephone and telegraph regulation yields no clearer indication of just how broad the class of communications common carrier was in the mind of Congress. How broad is the class, of course, has bearing on whether or not Packet is included. Two major reasons for such regulation were articulated during the first debates over communications regulation in 1910 and still seem to have bearing today. Senator Dixon argued in favor of regulation because:

> . . . the interstate telephone and telegraph companies are about the only remaining *public-service corporations* engaged in interstate commerce that are not under the control of the Interstate Commerce Commission [135]. (emphasis added).

In the House, Representative Bartlett stated another reason for congressional action:

> Why should not these necessary instrumentalities which citizens have to use, *which are monopolies in their particular lines of business*, be required to make reasonable charges . . . [136]? (emphasis added).

Passage of the Communications Act of 1934 was not intended to alter the underlying purpose of this earlier legislation. As Senator Dill remarked:

> The purpose of the proposed legislation is to make effective the power now written into the Interstate Commerce Act of control of the telephone and telegraph business in this country [137].

These legislative purposes are not mutually exclusive, but they tend in different directions in the context of the scope of FCC power and the question of Packet as a common carrier. On one hand, if the reason for regulating telephone and telegraph companies is to guard against monopolistic abuse [138] (a reason still alluded to by the FCC), then new offerings should be evaluated on the basis of their monopolistic propensities. And FCC power should be limited to regulating communications services which have the market characteristics that led to the regulation of telephone and telegraph companies in the first place.

The relatively small capital investment in Packet, its lack of control over its technology (which was developed originally by government research and is available to all), and other economic characteristics of the firm raise doubts in many observers as to whether PCI has any prospects of becoming a monopoly [139]. Indeed, this is also the case with most switching offerings or other services incidental to carriage as compared to the size, spread, and market power of the major communications common carriers.

On the other hand, if it is concluded that any organization which offers information transfer or distribution services via electrical media for hire is impressed with public interest and is therefore subject to FCC regulation [140], then the FCC has an extremely broad purview, in which Packet would be included. At the extreme, FCC regulation would include stock exchanges, travel services, research bureaus, answering services, and so on. Clearly, this is not the case. The House Conference Report accompanying the Communications Act to the floor noted that organizations such as news services (which then could turn away customers) were not to be included in FCC jurisdiction [141].

As new services develop and as functions incidental to carriage are perceived and identified, the question arises as to whether these offerings are to be treated as public service corporations, required to serve all comers, or whether these offerings should be allowed to turn away customers. Some decision must be made that an organization should be regulated because its service is impressed with the public interest or carried on for the public benefit. Generally such a decision is made today in the first instance by the legislature [142].

A service is not impressed with the public use or carried on for the public benefit because its offerer says so. Nor does the Communications Act seem to provide any mechanism for the FCC to establish new classes of public communications utilities. Arguably, the Commission's authority only extends to making decisions on whether or not an activity falls within the classes already established by the Communications Act.

And even the weight to be afforded to a Commission decision upon judicial review of such a matter is not a settled issue [143]. For example, the FCC seemingly had no mechanism to reach a fully binding decision on whether or not it had jurisdiction over cable television (CATV) services. After first announcing that it could not regulate CATV) [144], the Commission applied to Congress for legislation clearly empowering the FCC to do so [145]. In effect, the desired legislation would classify CATV services as public utilities; most logically, these services would be regulated by the FCC, since CATV technology supports a communications service. Congress did not act, so the Commission reversed its earlier holding and decided to regulate CATV services as "ancillary to broadcasting" [146]. This decision was appealed and finally affirmed by the United States Supreme Court [147]. The Court, of course, is clearly empowered to make or reverse findings that a service is within the scope of regulatory authority mandated by Congress [148].

Significantly for point-to-point communications services, the FCC selected its "other side," broadcasting, as the base from which to regulate CATV offerings [149]. Arguably, the FCC is bound in the point-to-point communications context to regulate only those services which perform and accept responsibility for the carriage of signals in a point-to-point system. Once an offeror operates technology designed to carry electrical signals from place to place and offers this service to the public generally, he qualifies as a common carrier and is regulated by the FCC. His carriage service exhibits specific monopolistic tendencies and each service he offers incidental to that carriage is also regulated. But services like Packet, which do not perform this carriage and only offer functions identifiable and separably from carriage, may not fall within FCC domain. And this may hold true even in the face of a purpose, announced in creating the FCC, of regulating commerce in communication seemingly without any such limitation.

The problem of the mismatch of the announced congressional purpose in creating the FCC and the Commission's actual power seems to surface particularly in the context of services relying on converging computer and communication technologies. Each area of examination of the scope of the concept of communications common carriage has led to plausible, yet inconsistent results when attempting to determine whether a service like Packet's should be allocated and regulated by the FCC. These inconsistencies seem to follow inevitably from the fact that Congress, in passing the Communications Act, seems to have established a framework for allocation of functions based on a static pattern of carriage and incidental services.

At the time of the passage of the Communications Act, only communications common carriers, the telephone and telegraph companies, offered signal carriage. But today computer use is capable of identifying services functionally and defining clearly the interface between signal carriage and services incidental, necessary, or essential to that carriage. Computer use thus has changed the pattern of carriage and incidental services from the structure implicit in the Communications Act. To a great degree, service offerors other than traditional carriers are now able to provide noncarriage services that might be called "incidental to carriage" under the older scheme.

The framework for allocation of point-to-point communication services made in 1934 does not seem to reflect the ways and means by which those services are being provided in 1974. Trying to force into that old framework new technologies and services will likely prove to be unsatisfactory [150] for the implementation of allocation policy based on clearly defined goals. The next section of this article discusses some of the options available to policy makers in deciding what structure will be used to allocate services such as Packet. The purpose is to point out issues and consequences that might follow from specific allocation decisions. While the section deals only with services incidental to carriage, it is designed to show that even a single allocation decision may have great impact on the price and availability of many other communications services.

V. OPTIONS AND ISSUES

Packet now has official status as a common carrier, certified as such by the FCC. But it is apparent that although the question of PCI's status has been answered, the issues raised by the application have not been settled, nor did the FCC really claim that it was settling them. The Memorandum Opinion granting PCI's application refers to both the technical and commercial volatility of the data transfer business [151].

The Opinion, however, only represents a view of the question taken from within the FCC. As we argued in a previous section of this article, the decision to employ the FCC's allocation mechanism need not have been made by the Commission, and indeed may not have been an appropriate decision for it to make. Congress and the courts also have procedures and doctrines concerning alteration of market structures, implementation of regulatory schemes, or imposition of liabilities. They too allocate functions.

In this section some of the options available to deal with the sort of issues raised by the PCI application are outlined. Much of the difficulty of these issues stems from changes in technology—specifically in services incidental to carriage—that have occurred since the Communications Act was passed in 1934. As the following section shows, the selection of a mechanism for allocation of communications functions can have a great impact on what communication services are offered, how they are provided, who gets to provide them, and how much users pay for them.

A. Option: An Unregulated Market

A number of arguments can be advanced for bypassing the FCC in the case of offerings like Packet's. First, Packet is not actually engaged in carrying signals. It purchases signal carriage from already regulated carriers. Packet simply adds a functionally identifiable data processing component to the carriage service available to all and then sells the whole package. Second, Packet's market characteristics are not those of the common servers who are specifically regulated by the FCC. And third, there has been to date no public showing by Packet that it intends to offer its service to the public generally nor any demonstrated need that it do so. Such a demonstration would be a prerequisite to regulation of Packet as a common server, even if it did offer a carriage service. In fact, Packet's service seems to be geared to meeting a very specialized demand. Packet may implement its service only where there is enough demand to support it, rather than offering the service to anyone who wants it or to the public generally [152].

At first glance, it seems reasonable to suggest that rather than allocating services like Packet's through the procedures of the FCC, such services might be allocated through the mechanism of the competitive market. Several issues would probably surface as a result.

1. Issue: Liberalization of the Prohibition of Resale of Carriage Service. At present, the resale prohibition incorporated into common carriage tariffs would stand as a barrier to allowing Packet to add computational facilities to a signal carrier's service and to sell the combined offering without restriction. Therefore, in order for Packet to be able to offer an unregulated service, the resale prohibition would need to be liberalized. In fact, a decision to allow Packet to offer an unregulated service might indicate that a common carrier should be required to provide service to any member of the public, no matter what the use—of course subject to network technical requirements and other legal constraints [153]. Such a decision would also begin to implement a policy of evaluating offerings functionally and regulating only those which actually carry signals from place to place.

Several consequences might follow the removal of the resale prohibition. First, under current tariffs, a customer or group of customers must purchase channel capacities in standardized amounts. The standards used are those of the carriers. It is rare for the standard offering of channel size to match precisely the needs of the customer. Customers may need only one-tenth of what they are forced to buy under the tariffs [154]. The excess capacity is frequently of no use to the customer, who presently is most often prohibited from reselling it. However, a proposed new AT&T tariff at least would allow customers to sell off excess capacity to other users, provided that no profit is made on the resale and that the customer has a direct interest in the communications transmitted [155].

Second, allowing resale of channel capacity might be expected to encourage competition in the market for equipment to be used for dividing up large capacity channels into smaller channels. This competition would include the carriers, who might find it necessary to meet the pace of innovation by developing advanced and efficient retail services themselves [156]. Third, the carriers might lose a good deal of revenue, while their customers would not need to pay for a great deal of unused, excess channel capacity. Western Union, for example, has estimated that removal of the resale prohibition would cost it $15 million [157]. The resale prohibition thus may be a tool of those common carriers who need it to protect their own economic vitality.

2. Issue: Increased Unbundling of Functions Incidental to Carriage. Leaving organizations like Packet to the unregulated market would indicate that such services should be offered competitively. Regulated carriers thus might be forced to continue to "unbundle" services incidental to carriage in order that competition in these services might be unimpaired. Regulations in line with the maximum separation principle of the Computer Rules would need to be implemented to prohibit carrier cross-subsidization of prices for services in competitive markets. In other words, carriers will need to be prohibited from undercutting competitors in incidental services by subsidizing artifically low prices for competitive services with profits from monopoly services.

The effects of such a ban on cross-subsidization, or cost averaging, would have widespread effects. For example, unregulated "brokers" who divide the carriers' standard-size channels into smaller capacity channels could insure that the price for dividing up large channels accurately reflects the cost of doing so [158]. But this trend toward true cost pricing might also mean that consumers will pay more for some noncompetitive services, such as long distance directory assistance or rural telephone, whose prices below cost are traditionally subsidized by monopoly profits from other services.

3. Issue: Solidification of Carrier Market Position. A competitive market for services that are incidental to carriage raises questions about the dependence of these services on established voice and record carriers. Competitive directory assistance would be useless without telephone numbers. Automatic Location Identification needs the telephone system in order to operate. Packet's nonmonopolistic economic characteristics depend on Packet's ability to purchase services from natural monopolies. Do these facts indicate a solidification of the monopoly position of those already participating in closed carrier markets? Or do they indicate a desirable function separation of monopolistic and competitive services?

Whether or not continued introduction of Packet-type services will generate increased demand for carrier services that will benefit those already operating in carrier markets is a question deserving careful study. The question implies that selection of the competitive market for allocation of Packet-type services may involve policy decisions concerning how much control over communications channels is to be vested in a few signal carriers.

4. Issue: Availability and Effectiveness of Other Allocation Mechanisms. A decision that Packet should not be subject to FCC regulation raises questions about the role of the FCC as allocator of functions in the communications market. Such a decision would suggest that the FCC should deal only with offerings that employ technologies and assume the responsibility for carrying electrical signals from one place to another. Wherever possible, functionally identifiable, incidental services should be evaluated independently, and not necessarily by the FCC.

It has been assumed in this section that allocation of these incidental-to-carriage services will take place through the competitive market. It is entirely possible, however, that this system for allocating services and functions may prove unsatisfactory. Yet, even in the face of unsatisfactory performance in the competitive market, it may still be decided not to employ the allocation procedures of the FCC. Other systems could be called upon to establish procedures for deciding who is to offer what services, and how, in the market for Packet-type services [159]. A crucial policy question is whether or not these allocators wish to be available, or indeed are competent to deal with such a dynamically changing market.

B. Option: Use the FCC Allocation Mechanism

Several suggestions have been made as to why the FCC should regulate services such as the one offered by Packet. The most notable of these suggestions, put forth by the FCC itself, is that the Commission was established to regulate all communication services:

> [T]here is a *bona fide* question as to whether . . . decisions may be construed to permit the Commission, by the exercise of administrative discretion, to exempt from the provisions of the Act applicable to interstate common carrier communication services those activities which clearly or admittedly constitute a public offering of communications services. This is because Congress has specifically ordained that all such activities are *subject* to regulation and has fashioned a specific scheme of regulation therefor [160].

Some of the issues raised by FCC regulatory control over all communication services for hire, whether or not they involve carriage, seem to be readily apparent.

1. Issue: Role of the FCC. The FCC may not have the regulatory tools to deal with a market that does not resemble that of the common carriers. Close examination of the statement of the Commission above indicates that, in following its interpretation of its congressional mandate, the FCC may be applying to *all* public communications services the statutory provisions applicable specifically to common carriers. "Public communication services" may include offerings such as Packet which arguably are not conventional signal carriers nor behave like signal carriers. The result of this statutory application may be the development of a regulatory scheme not at all suited to the Packet service market. How responsive can the FCC be to the task of allocation and regulation in such a different and changing context?

2. Issue: Innovation in Point-to-Point Communications. The market for services like Packet, based on computer technologies, has been characterized by rapid technological development. Those organizations already regulated, however, may desire to slow implementation of new services and technologies. With any such pressure, a procedure for subjecting new, computer-based, noncarriage offerings to preliminary scrutiny by a regulatory agency might increase risk to competitive market entry and slow innovation and implementation in an otherwise dynamic market.

3. Issue: Regulatory Use of the Resale Prohibition. The FCC has called for a case-by-case evaluation of new communications offerings. This evaluation is to determine what each new offering is: data processing or common carriage, hybrid data processing or hybrid communications. Yet after *GTE* [161] it is clear that the FCC does not have general jurisdiction throughout the range of these services. One feasible way for the FCC to determine that a party subject to its case-by-case examination is not within its jurisdiction, hence properly not subject to its powers, is to have that party complain to the Commission. A common carrier's refusal, on the ground that the new service is reselling carriage in violation of tariff rules, provides just such a mechanism.

The FCC may need to rely heavily on traditional common carriers to police the border between data processing and common carriage, hybrid data processing and hybrid communication. As the Commission does so, the carriers' policing tool, the resale prohibition, may shift from mere corporate practice to an important regulatory tool. There are several issues that would seem to result from such a shift. How would the role of the FCC change vis-à-vis those it regulates?

Would the FCC be able to guide the proliferation and provision of carrier and noncarrier services in disinterested pursuit of the public interest? Or would carrier interests unavoidably grow to be "the public interest"? These are only some of the policy implications that should be carefully considered before the resale prohibition is systematically employed as a regulatory tool.

C. Options for the FCC

Policy considerations, of course, might lead to the conclusion that Packet and other services like it should be allocated and regulated by the FCC. Such a conclusion might be reached through a number of routes. It might be decided that all such services should be regulated as public utilities by the expert regulatory body in the communications area, the FCC. Policy might dictate that the vitality of existing common carriers should be guaranteed and that the resale prohibition should be employed to achieve this end. Or, it may be desirable to reduce or keep low prices for some costly carriage services and to allow carriers to earn monopoly profits in non–carriage protected markets in order to accomplish that result. The FCC, in turn, seems to have several readily apparent options to deal with Packet and other such offerings.

1. Packet as "Ancillary to Carriage." The FCC might eventually decide that Packet does not fit into the model of a common carrier or even a hybrid communications offeror and should therefore come under FCC power in some new category of FCC regulation. Packet and others like it could be classed as "ancillary to carriage," repeating the FCC's decision to make CATV services "ancillary to broadcasting," and presumably repeating the lengthy rule making procedures followed in the CATV case.

2. Packet as a Connecting Carrier. Having concluded that PCI should be a common server, open to the public generally, the FCC might still in the future decide that since Packet actually only connects channels, it should be regulated as a connecting carrier [162], subject only to §§201–205 of the Communications Act [163]. This would exempt Packet from the Section 214 procedures governing the construction, acquisition, and operation of lines [164]. While such a decision would nominally recognize that Packet is not actually carrying signals, it would still amount to a finding that PCI's service is carriage.

Any finding that PCI performs carriage expands the meaning of communications common carriage, with at least two immediately foreseeable consequences. First, the Bell System benefits from a declaration that Packet's service is carriage. AT&T is operating under an antitrust consent decree which prohibits it from engaging in almost any business that is not common carriage [165]. Therefore, any new service dubbed "common carriage" by the FCC can also be offered by Bell under the consent decree, expanding Bell's role in the market. Second, classifying PCI's service as carriage eliminates the need to liberalize the prohibition on resale of carriage capacity. Under present practice, classified as any sort of carrier, PCI is free to contract—without regulatory interference—to send business over leased lines for hire.

3. Packet as a Section 214 Common Carrier. Finally, we come to the option chosen by the FCC. The benefits to established carriers discussed immediately above carry over to this option. The conclusion that PCI is a common carrier possibly indicates an expansion of the definition of common carrier to include anyone whose package offering includes a carriage service, whether or not the offeror provides the carriage. The Bell System and other line operators benefit from the new business without losing the protection of the resale prohibition.

More important, however, is the fact that the FCC, by its decision on Packet, seems to have bootstrapped itself into an assertion of discretionary power over any commercial information service using wire, cable, or radio channels for foreign or interstate point-to-point communications. This assertion of power, if sustained, could make the FCC the allocator of information functions using electrical communications channels, or at least could give the Commission the power to chose whether or not to regulate any particular information function using those channels. And this grant of power to the FCC by the FCC seems to be considerably greater than the power granted the FCC by Congress in 1934. Arguably the Communications Act does not appear to empower the Commission to regulate identifiable noncarriage functions throughout the point-to-point communications industry, but rather only empowers the Commission to allocate the special functions defined by the technology of carrying signals from place to place.

VI. CONCLUSION

As new functions are perceived and new services are developed, the natural tendency is to view and include them within the established technological and institutional environment. We have seen how this tendency puts great pressure on the concept of communications common carriage, perhaps expanding it beyond the scope intended by those who developed it. A concrete manifestation of this expansion of definition is the expansion of the power of the FCC from the regulation and allocation of a few organizations operating specialized technologies, to the regulation and allocation of functions throughout a much greater range of the market for communications and communications related functions and services.

Since information transfer and distribution are vital social functions, then the patterns of communications established play an important role in structuring our society. Thus, the more services and functions included within the purview of the FCC, the more power and control over many aspects of life in our society are vested in that administrative agency. Given a wider range of services to regulate and allocate, the FCC has wider powers about which services to encourage, promote, or have subsidized, and which services to discourage, place in a subordinate role, or have taxed.

Some of the questions thus raised are: What criteria are employed in making these political decisions? What constituency is the FCC serving? Is the FCC the proper locus for this broad allocation power? Is there a basis in law for this expansion of the FCC?

Part Three

Foreign Policy: The High and Low Politics of Telecommunications

William H. Read

TELECOMMUNICATIONS AND INTERDEPENDENCE

Global interdependence—of which telecommunications is the very nerve system—is widely acknowledged as a salient trait of the new world order [1]. Control over telecommunications thus assumes a new place on the foreign policy agenda. Yet, is the established policy process still adequate? As Figure 3–1 shows, something more than technological advances and cost cutting are at hand: the rise of international telecommunications reflects the growth of so-called "transnational" activities engaged in by border-spanning organizations such as multinational enterprises, cross-cultural affinity groups, and even military commands.

The changed condition of international telecommunications not only mirrors how much these transnational activities are expanding, and thereby further strengthening interdependent relationships, but also how reliant on the global telecommunications infrastructure those relationships have become, and furthermore how telecommunications policy looms as an instrument for shaping interdependence. As yet there is no crisis which compels restructuring of the international telecommunications policy making process. But, because transnational activities are now far more than casually reliant on international telecommunications, it seems prudent to reevaluate the process by which telecommunications foreign policy is made before something like a "Law of Communications Conference" forces the issue.

In the bygone era when the principal issues were technical standards and tariffs, the interests and therefore the influence of carriers were predominant; today, however, the primary stakes are broader. The benefits of abundant and versatile telecommunications have created information flows within transnational organizations which can be, as one insider has put it, critical . . . imperative . . . vital. And changing conditions have also given birth to noncarrier service suppliers, notably in the computer/communications field. Still the commanding actors continue to be carriers. While nothing suggests that they should now take a back seat in the making of bilateral and multilateral telecommunications policies, is the absence of diverse inputs into the policy process unnecessarily risky at a time when telecommunications has become one of the international life lines of global interdependence—life lines which strategically are both a strength and a weakness?

All international activity is communications-dependent to some degree. The missionaries of old relied on their printed bibles, while the essence of, say, international broadcasting is the ability to communicate into foreign countries with radio transmitters. What concerns us here is the fact that technological breakthroughs in telecommunications, fostered by World War II and developed ever more fruitfully since, have catalyzed this virtual explosion in both the depth and kinds of international exchange. So rapidly did various forms of exchange grow after the war that statesmen began to speak of a "family of nations" and Marshall McLuhan coined the phrase "global village." Both phrases reflected advances in international telecommunications, which have developed so rapidly that the vocabulary of statesmen has since been upgraded to "interdependence," while that of political scientists now features terms like "interconnectivity" to describe the emerging social condition of the world [2].

What gives rise to these labels is the proliferation of transnational activity. One way to explain what is meant by the word "transnational" is to say that it is analogous to what has been called transstate activities in the United States [3]. Examples abound—development of interstate commerce, the rise of national amateur sports competition, and the ascendency of the national armed forces, to cite a few. Today we see that these and other national activities have international manifestations—multinational corporations, the modern Olympics, and overseas military commands. Hence the word transnational, a word that relates to activities which are imperfect fits for the geographical divisions of countries. Transnationalists—a not wholly American group of international actors—speak of the scope of their operations in terms of groups, markets, or commands.

This ability to use functional as distinct from territorial terminology is enhanced to the extent that telecommunications disregards international borders. It is not merely that technology enables information to be conveyed near instantaneously across great distances, but also that messages that travel by international telecommunications are less adaptable to selective control devices such as passports and customs clearances. It is not surprising then that the reasons why America plays the leading role in transnational order include both U.S. leadership in telecommunications applications and U.S. activism in championing the principle of free flow of information across national boundaries.

CHANGING COMMUNICATIONS CONDITIONS

In the past, carriers provided both telecommunications circuits and services (a distinction of some importance, as will be seen). A few carriers are private companies, like AT&T, or quasi-public, like COMSAT, but most are specialized government agencies of the Post, Telegraph, and Telephone (PTT) variety. It is these carriers (known internationally as administrations) who effectively negotiated bilateral operating agreements and played leading roles in the multilateral forums of the International Telecommunications Union (ITU) and the International Telecommunications Satellite Organization (INTELSAT). Thus carriers have controlled both the actual telecommunications systems from end to end and the political process which decides questions of interconnection and rates for what was until recently a comparatively small amount of international service.

In recent years the nature of telecommunications has evolved under U.S. leadership from a period of scarcity and limited flexibility to one of abundance and versatility. Space satellites and high capacity transistorized undersea cables have greatly increased the number and quality of circuits, while electronic technology also has created an array of communications choices. In response to these developments, it is common today for major American organizations no longer to rely exclusively on carriers for service. The role of carriers often is reduced to only leasing circuits to organizations which operate their own private service networks or to new service suppliers such as the entrepreneurs who are selling computer services via telecommunications. An organizational response to this situation has been to create the position of communications manager, an individual whose task is to see to it that services ranging from plain old telephone to interactive computer/communications be tailor-made for his organization—globally, if need be.

Even so, a skeptic—especially an American who has become accustomed to a generally well working telephone system—might legitimately ask whether the impact of technology has any significance beyond eroding carrier control over the telecommunications system? One answer is that policies can be politically and economically potent. The situation in Europe today presents an interesting hypothesis. In response to transnational organizations either establishing their own data communications networks or buying data communications services from private companies, the Consortium of European Post and Telecommunications Administrations (CEPT) is working towards harmonization of their individual data networks. If achieved, harmonization would be another step toward European political integration.

A far more significant idea being discussed in some circles is to make the European Community responsible for "long distance" telecommunications—i.e., the Community would provide for its member countries the type of long lines service offered by AT&T in the continental United States. The idea is clearly one of the most advanced federalism schemes for Europe. It could promote continental cohesion in a vital sector and thereby be a remarkable agent of political integration. Moreover, a European long lines entity would have the power to effectively favor European industries over others. One implication could be that America's communications-dependent multinational corporations could be hamstrung by telecommunications policies that eroded the efficiencies of their operations and thereby made them less competitive and thence vulnerable to otherwise less efficient European firms.

This scenario is not presented as being inevitable, but rather to suggest the potency of telecommunications policy. Yet there is a very limited appreciation that telecommunications policy has an inherent power to shape and influence other activities. This is not to say that there is no perceived need for viable systems to provide reliable service at a relatively low price. Indeed, most of the policy action, domestic as well as international, occurs in this realm. Witness, for instance, continuing public policy debates within the United States over whether the benefits of introducing carrier competition outweigh the risks of eroding the basic integrity of the public telephone network run nationally by AT&T. Abroad, similar concerns have been echoed over whether carriers should retain exclusive control over all international services.

What triggered policy debates in both instances were electronic technological breakthroughs which moved telecommunications into its era of abundance and versatility. The wide availability of technology operated as a coercive fact internationally, for it brought about a rather reluctant consensus among national carriers to relinquish several degrees of control. Despite some lingering misgivings that they might be cutting their own throats, the carriers in the late 1960s approved recommendations within the telecommunications wing of the ITU [4] on lowering two nontechnical barriers. To be sure, liberalizing tariffs on "alternate voice data" circuits and permitting a leased circuit user to derive additional low speed teletypewriter channels from a circuit may not sound like the stuff of high international politics. Indeed, at one level it may be said that nothing was at stake but that carriers had become too comfortable both with their exclusive control over hitherto scarce telecommunications circuits and with their technically outdated institutional distinctions between telephone and telegraph.

But there are economic and political ramifications of broader scope. What the recommendations legitimized, for telecommunications administrations willing to accept them, was a right to allow all transnational organizations to do what a few—notably the U.S. military and the airlines communications society—already were permitted to do: configure private networks internationally to meet their own specific service requirements. A door to even greater operational efficiency by transnational organizations had been opened, and those who walk through are strengthened as agents of interdependent world order.

One of the more impressive illustrations of what is occurring is the availability of computer time sharing on a global basis via private networks. This form of international communications has raised some policy issues in its own right: the man versus machine privacy question on an international scale, and a data haven versus national sovereignty question. To some extent these issues are being addressed, notably by the Organization for Economic Cooperation and Development. These are not unimportant concerns, but they are nonetheless ancillary to the principal question of whether international telecommunications as the very nerve system of global interdependence is being guided by an adequate policy mechanism.

THE POLICY PROCESS

Although the conditions of international telecommunications have changed radically in recent years, the carrier-dominated policy process has not. Does this suggest that transnational organizations should attempt to influence the workings of the ITU, INTELSAT, or CEPT as some of the new service suppliers already are doing? If so, how? Through individual or collective action? And what institutional responses, if any, should the carriers make? And further, what larger role, if any, should governments play?

Thoughtful answers are needed in order to make the policy framework relevant. In attempting answers, however, three considerations are worth being mindful of. First, there is no crisis. A well established axiom of international affairs might be called the immediacy rule, which in practice means that changes are not undertaken absent an immediate problem. The hope, however, should be to avoid the necessity of having to rely on crisis management at some level of deterioration in international telecommunications and, instead, now pursue the less dramatic work of scanning and planning—a task which is coming to be seen as being of considerable importance in an interdependent world [5].

There is limited recognition of this need among some telecommunications specialists, but without at least an adequate dialogue, which includes inputs by organizations whose reliance on telecommunications is critical, can there be meaningful scanning and planning? Even getting this process started and then institutionalized may well be but a first step in bringing the policy process into focus. Structural changes for policy implementation also may be required. Merely creating a better arrangement to foresee consequences risks in telecommunications the same neglect that the Johnson White House made of food even though its own science staff had spotted that impending world problem [6].

The second consideration is that introducing transnational organizations into the policy process brings to the fore (and thereby risks exacerbating) a thorny question: where does interdependence end and interference begin? Since communications policy is traditionally viewed as a primary instrument for exercising national sovereignty, the suggestion to bring transnational actors onto the policy stage is likely to meet resistance. But assuming that factors favoring pursuit of interdependence outweigh those supporting nationalism, then the management of interdependent relations should be facilitated by the politics of inclusion. Otherwise the situation becomes fraught with unnecessary risks of ill informed, narrow policy formulation that give rise to unanticipated outcomes.

Already there is incipient awareness of these risks. At the last plenipotentiary conference of the ITU a representative of the U.S. Defense Department, the world's biggest telecommunications user organization, was included in the American delegation. And in the nongovernmental sector a tangible sign of awareness was the recent formation of the International Telecommunications Users Group (INTUG), whose members, like the U.S. military, appreciate that they too must participate in telecommunications policy making because for them the stakes are very high.

The third consideration is political. The process now must take account of the fact that the United States occupies a less imposing position in world affairs. After World War II, American transnationals flourished because the United States had the political power to keep the international system open (i.e., to assure freedom of operation), and transnational organizations had the necessary means of transportation and communication to pursue strategies abroad which are rooted in the American experience of creating national enterprises within a federal system of states. But while technology has offered improved means (jet liners, computers, satellites), the ability of the United States to keep the system open—to enforce the general concept of freedom to operate and the specific principle of free flow of information by any medium—has been challenged.

A crucial question then is this: Will the concept of freedom to communicate endure the current phase of international affairs in which the United States is experiencing a relative decline in its international political power?

Updating the U.S. Policy Process

It can be argued that as a matter of national policy the United States should not seek to preserve an international environment conducive to transnational activity. This view may be especially strong among those who are disenchanted with multinational corporations. But there is a danger here of throwing the baby out with the bath water. Transnationalism is an American style of operation of which multinational corporations are but one form. Our concept of transnationalism is rooted in the American experience of what previously was termed transstatism and is embodied in such U.S. constitutional principles as free speech, full faith and credit, and due process. Moreover, the other industrialized democracies also are acquiring considerable stakes in transnational activities. At heart then is a commitment to a system of world order in which this style of activity is safeguarded. The trick is to devise procedures for honoring that commitment effectively with arguments, since it can no longer be asserted as a byproduct of U.S. power.

Both foreign critics and domestic neo-isolationists may fault this reasoning on grounds that it is, in effect, mere justification for further arrogant attempts to impose American influence and values abroad. However, since power politics is of diminished value, greater reliance has to be placed on the more difficult method of persuasion. But to employ this tactic with reasonable prospects of success, transnational organizations will have to enter into the telecommunications policy making arena because the politics must now of necessity be that of joint gains, not one of trade-off. In other words, countries will have to become convinced that the communications needs of transnational organizations should be satisfied so that host countries will realize benefits from their presence there. The old politics of host countries trading off assured communications for American transnational organizations in return for being allowed to stand under the protective U.S. security umbrella are of less value now.

The suggestion here is not for coercion, but for persuasion. And the assumption here is that the nations of the world—the United States included—in pursuing their self-interests rely on the continuation of a widespread and deep American involvement in world affairs. It is within this context that the future of international telecommunications must be considered. At minimum this requires arrangements whereby the views of the diverse transnational organizations are made known and are taken into account in arriving at policy decisions. No recommendation is made here for how this ought to be accomplished. But it needs to be recognized that so far neither has the traditional International Telecommunications Union been adequate to this task (for it has remained essentially a "carrier's club"), nor has the Organization for Economic Cooperation and Development filled the gap with any of its recent undertakings in the information field because there too transnational organizations have not participated.

To neglect their pragmatic interests—or worse, to presume to know their interests—is an imprudent risk at a juncture when technology has enriched international telecommunication service. Already there are some visions of how electronic communications may be a promising tool to help better organize interdependent relationships. But there are also sharp political disagreements over control of advanced information technologies.

Some examples: Recent world monetary problems gave impetus to the idea of an international electronic funds transfer network; ways to implement such a scheme, however, divided the banking and carrier communities. And even though a compromise agreement was reached, continuing troubles are anticipated. In Sweden, credit cards accepted around the world are fine, so long as no data is communicated abroad without a license. In Britain, the post office's reaction to computers that can pump information through circuits faster than teletypewriters was a surcharge. At the ITU, the official position on computer communication is, predictably, only-at-the-discretion-of-carriers, while in Ottawa, consideration of that subject centers on a for-Canadians-only policy.

As the illustrations show, the basic issue is control; and the stakes are high, since interdependence is fostered by transnational organizations which rely on no longer scarce or inflexible international telecommunications. The carriers, since they are accustomed to being the main policy players, must perceive this all too well. There is evidence too that transnational organizations and government specialists are beginning to awaken to the fact that telecommunications, driven by its technology, is emerging as a vital resource in international relations. It need not emerge as a crisis.

As yet the level of tensions is relatively low, with the possible exception of Canada (telecommunications issues there are part of a festering "information" problem which seriously affects Canadian–U.S. relations). If tensions are to be prevented from escalating, however, the policy process needs to be brought into proper focus. This may require a reevaluation of the roles of both carriers and governments; it certainly requires that steps be taken to achieve meaningful policy participation by those transnational organizations which rely on telecommunications.

There unfortunately will be enough implacable problems to deal with in a world of interdependent relationships without allowing telecommunications—the very nerve system of interdependence—to become one of them. In sum, the issue is how better to organize the process by which international telecommunications policy is made. To further support this view, a more detailed analysis is offered of the two premises, viz. as both the world and technology have changed, so have the stakes in telecommunications.

A TRANSNATIONAL WORLD

Today the United States is in the midst of some fundamental changes with regard to its relationship to the rest of the world. Already a domestic consensus has developed that the U.S. neither can nor should be the world's policeman or its banker. Such feelings of neo-isolation are, however, checked by an appreciation that this country has inextricably involved itself with other countries. The label for this involvement is "interdependence." World commerce offers numerous examples, some of which are highly visible. From Iran the U.S. buys supplies of oil, which in turn enables the Shah to buy American know-how and goods with which to modernize Persia. Even

America's bastion of autarky, the Pentagon, no longer considers military procurements abroad absolutely unacceptable.

The condition of interdependence, or for that matter, either dependence or independence, rests not merely on an evaluation of any country's own resources vis-à-vis its perceived needs; what is also required is a process of actually achieving balances, as in the case of Iran and the United States. During this time of international change for the U.S. (and other countries as well), it is therefore relevant to ask what processes are at hand through which change occurs—and indeed, do the processes themselves influence change?

While traditional processes remain with us and in at least one memorable incident facilitated a dramatic change (when Henry Kissinger secretly traveled to Peking, he filled the classic diplomatic role of a plenipotentiary), another process by which many of America's relationships with the rest of the world are conducted is contemporary and particularly—although not exclusively—American. The process is called transnationalism, and it flourishes principally because modern technology has presented hitherto undreamed of means of international communications and transportation. The significance of transnational activity may be seen first in comparison to the traditional process of international relations and second in terms of the organizations which engage in those activities.

Transnational Order

At their beginning, Americans concerned themselves with world order. Dissatisfaction with colonialism was, after all, what prompted the American Revolution. But from then until this century, the U.S. had minor influence upon the way in which the world was organized, the Monroe Doctrine being a notable exception. The events of the twentieth century, culminating with U.S. victory in World War II, put the United States in a position to indulge its global preferences for nationalism (the end of colonialism) and internationalism (the founding of the United Nations system).

The ascendency of the United States in world affairs also fostered transnationalism, a less predictable preference, although one that had firm roots in both the American federal experience and another notable nineteenth century foreign policy pronouncement, the Open Door (to China) policy. The implications of this preference were not quickly perceived, even by Americans. The postwar focus was on power politics, a view that has persisted with validity to this day. The power politics model was well summarized by Joseph S. Nye, Jr. and Robert O. Keohane in the introduction to their book, *Transnational Relations and World Politics:*

> Students and practitioners of international politics have traditionally concentrated their attention on relationships between states. The state, regarded as an actor with purposes and power, is the basic unit of action; its main agents are the diplomat and soldier. The interplay of governmental policies yields the pattern of behavior that students of international politics attempt to understand and that practitioners attempt to adjust to or control. Since force, violence, and threats thereof are at the core of this interplay, the struggle for power, whether as end or necessary means, is the distinguishing mark of politics among nations. Most political scientists and many diplomats seem to accept this view of reality, and a state-centric view of world affairs prevails [7].

The problem with the "state-centric view," according to Nye and Keohane, is that "[a] good deal of intersocietal intercourse, with significant political importance, takes place without government control." They cite as examples trade, personal contact, and communication among western countries. Moreover, the power politics model is associated with traditional activities of central governments in the conduct of their foreign relations. Accordingly, based on a reading of history, one would have expected the United States to have acquired an empire at the end of World War II. Indeed, some statesmen and scholars have argued that that is precisely what did occur.

But if America gained an empire, it certainly was neither of the ancient Roman nor the more recent British variety. For the postwar "empire" of the United States is without territory, the occupations of Japan and part of Germany having been ended in historically short time periods. Yet an American empire (if that be the correct term) of functions did come into being. It has not, however, been predicated on control over, but rather access to, foreign lands. This, it may be remembered, was what the Open Door policy was about. In 1899 the United States did not seek, as did European powers, extraterritorial rights in China. What the U.S. sought was an understanding that all nations have equal trade access to China.

Transnationalism, then, embodies two concepts: one is an enlarged scope of international activity due to the significant roles being played in world affairs by nongovernmental organizations, the second a concept of significant international activity (either by government or private organizations) which is aimed not at securing control over another's territory but having access to it. The qualifying word significant has been employed, for the suggestion here is not that either concept is new (the Catholic church has been practicing transnationalism for centuries), but that these transnational activities are beyond inconsequential levels and their numbers far exceed similar pre-World War II activities.

Transnational Organization

As was shown in Figure 3–1, the number of international organizations that are binding peoples of the world together has more than doubled in a dozen years. Like-minded individuals in fields as different as agriculture and literature, health and religion have been joining an expanding number of international affinity groups. Telecommunications has helped make the rise of these and other transnational organizations possible. Indeed, Samuel P. Huntington has written that since World War II the world has experienced a "transnational organizational revolution." He made that statement in an article that began with the following list of organizations and commentary:

Anaconda	J. Walter Thompson	Ford Foundation
Intelsat	Air France	Catholic Church
Chase Manhattan	Strategic Air Command	CIA
AID	Unilever	World Bank

> These twelve organizations appear to have little in common . . . Yet they do share three characteristics. First, each is a relatively large, hierarchically organized, *centrally directed bureaucracy* (emphasis added). Second, each performs a set of relatively limited, specialized, and, in some sense, technical functions. . . . Third, each organization performs its functions across one or more international boundaries and, insofar as is possible, in relative disregard of those boundaries. They are, in short, *transnational organizations*, and the activities in which they engage are *transnational operations* . . . The increase in the number, size, scope and variety of transnational organizations after World War II makes it possible, useful, and sensible to speak of a *transnational organizational revolution* in world politics [8].

The very first point Huntington makes about these organizations is relevant: "each is a . . . centrally directed bureaucracy." Suppose for a moment that each was denied its modern means of international communications. To what extent then could any of their bureaucracies be centrally directed? Indeed, how many of these organizations would attempt transnational operations if this were still the age of communications scarcity? Of course, the answer would depend on individual case studies, but a reasonable assumption would be that all (save perhaps for the Catholic church) are today telecommunications-dependent in their transnational activities. Put another way, it is difficult to imagine any of these organizations (except the Church) being able to mount and sustain over time a transnational operation that was independent of assured modern communications.

How was this assurance gained? It was a vital part of what may be called access accords—i.e., agreements either formal or implied

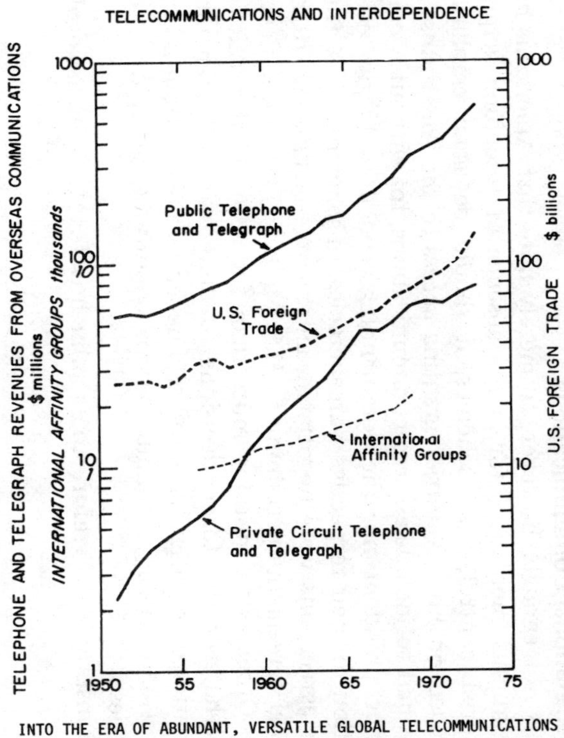

Figure 3–1. Into the Era of Abundant, Versatile Global Telecommunications

1950: Overseas messages to and from the U.S. total 773,000. A quarter-century later the number was about 38,000,000.

1956: The first transoceanic telephone service by cable begins. Two decades later a sixth transatlantic submarine cable—with a 4,000-circuit capacity—comes into service. Meanwhile, in a single decade the capacity of international satellite communications jumped from 150 to over 10,000 circuits.

1964: The international telecommunications society of airlines starts a computer communications network which a decade later is sending some 300,000,000 messages a year around the world at an average cost of five cents a message.

1969: AUTOVON, the Pentagon's domestic network, goes worldwide, presaging a proliferation of private international networks as carriers become more willing to lease their no longer scarce supply of circuits.

1970: Another inroad into the traditional notion of end-to-end service provided exclusively by common carriers occurs when General Electric links European customers to a Cleveland computer "super" center by satellite; truly global computer time sharing follows as GE spreads its service through eighteen time zones.

1975: Direct dialing from New York to London is priced at $3.60 for three minutes, with a higher quality connection that the booked-in-advance radiophone, which once cost as much as $75.

1976: Over 250 banks in North America and West Europe begin to use SWIFT, a network of computer communications lifelines that soon will handle 300,000 daily financial transactions among the industrialized democracies.

which allowed American multinationals sufficient freedom to operate in foreign countries. Typically, this freedom permitted regional or global strategies patterned after domestic experiences. During the nation's second hundred years, American business grew beyond the concept of individual state markets loosely linked in a federal scheme. Economically the country had become integrated into regional and national markets, while the business strategies which had accomplished this had been inculcated into large corporations. The trend in government went in the same direction as evidenced by, for instance, the dominance of the national armed forces vis-à-vis state militia.

So when incipient transnationals crossed the Atlantic they were conditioned to tap the "European market" or to defend the "European continent." The vice president for international operations of a manufacturing company was prepared to have components made in two different European countries and assembled in a third. The commander of American air forces in Europe was similarly prepared to base his war planes in Germany, supply them from France and maintain them in England. Yet, these were uncomfortable notions to some non-Americans. As J.J. Servan-Schreiber once put it:

> The American Challenge is not basically industrial or financial. It is, above all, a challenge to our intellectual creativity and our ability to turn ideas into practice. We should have the courage to recognize that our political and mental constructs—our very culture—are being pushed back by this irresistible force [9].

Transnational Communications

It is not enough, however, to merely note that Americans had a transnational mentality, for in the absence of forcefully acquiring territorial control, their schemes, it should be again emphasized, depended on host countries granting *access* to the Americans. And access had to include assured communications, for without it neither the vice president nor the commanding general could give central direction to their respective bureaucracies. In some instances, formal access agreements were negotiated, an example being some 70 international agreements pertaining to U.S. Military Telecommunications [10]. Others were nothing more than the absence of restrictions. But, whatever the form, decisions to undertake and then pursue transnational activities have and will depend on perceptions of terms of access, including communications conditions, granted by host countries.

It is instructive, therefore, to examine how the situation has evolved since American transnational activity began to flourish after World War II. Of particular interest here is a comparison between periods of communications scarcity and communications abundance, with 1965 being a reasonable dividing line between the periods. Prior to 1965, the United States actively and successfully pursued a policy of political expansion into world affairs and this coupled with America's command over technology produced an environment abroad conducive to the granting of highly favorable access agreements by host countries to U.S. transnational enterprises. The communications aspects of such agreements were, moreover, advantageous to host countries for they generated otherwise unanticipated sources of revenue based on international communications tariffs, which traditionally had been higher than domestic. Reinforcement of this situation was provided by the International Telecommunications Union, a United Nations body which, like its parent organization, then came under the domination of the United States and its allies.

Since 1965, however, the United States worldwide political power has eroded relative to that of other countries. In this environment, host countries have been adjusting once generous terms of access—and it is happening virtually everywhere. Greece has renegotiated the status of U.S. military forces agreements, the OPEC countries have turned the American oil companies into tax collectors, Canada has repealed tax measure beneficial to *Time*, and so on. Meanwhile, a key dynamic upon which transnationalism prospers—communications—has boomed, and with advances in information technologies come even greater capabilities for transnational operations.

But the extent to which transnationals will be allowed to enhance their operational abilities by using new information technologies is very much open to question at this time. The case of cross-border data flows exemplifies the situation. Why, asks the Canadian nationalist, should U.S. multinationals be allowed to exert even more influence over Canadians through the use of cross-border computer/communications networks? Why, asks the European PTT official who has come to depend on now sizeable revenues from international Telex communications, should he authorize a new service which he believes would reduce that income? Why, ask Third World delegates who have the votes to control the ITU, should they support western positions that help the multinationals?

Answers are not clear in part because host countries no longer see a trade-off between granting favorable access to American enterprises and being allowed to stand under the protective U.S. military umbrella as being as advantageous as it once was. Moreover, nearly every country today—with Japan a possible exception because of its homogeneous society—experiences internal tensions between the compulsion to be modern and the desire to safeguard its heritage. The issue is the economic benefits of modernization (read "Americanization") at the price of surrendering part of one's national identity. For a Canadian, for instance, the conundrum can be between wanting to have the benefits offered by an American consumer credit card organization but not wanting a dossier on him stored in an American computerized data bank. It is within this political setting that telecommunications issues affecting transnational organizations are addressed.

NEW COMMUNICATIONS CONDITIONS

Until the late 1960s, a transnational organization satisfied its telecommunications requirements on an ad hoc point-to-point basis either through the switched systems available to the general public or by leasing from a carrier a single purpose line for the organization's sole use between two locations. But electronic technology has made much more than this feasible and attractive to organizations. Many—notably the military, the travel and information industries, and a growing number of other organizations as well—now operate their own international networks for multiple types of traffic. Some even do cross-border computer communications, although many choose to buy this service from General Electric, Control Data, Tymshare, and others who are distinct from the traditional carriers.

Whereas in the past carriers exclusively offered end-to-end services that were functionally different (telephone versus telegraph), today the carriers often provide only the intermediate circuit facilities and the organization provides its own services. And in so doing, organizations tend to disregard the no longer technically viable distinctions between voice and record traffic. The telecommunications industry, however, more or less still adheres to service separations for nontechnical reasons. Indeed COMSAT, acting as a "carriers' carrier," wholesales technically identical transoceanic satellite circuits to middlemen U.S. carriers, who then categorize them by service at retail. This arrangement is a matter of approved public policy (in this instance by the U.S. Federal Communications Commission) and it exemplifies what telecommunications policy making normally is about.

USERS—PAST AND PRESENT

Policies are said to be made in the interest of telecommunications users—a word which in recent years has acquired added meaning for the policy process. Historically, neither the individual telephone caller nor an organization which operated its own switchboards has been considered a customer or client in the ordinary business sense, since telecommunications generally is outside the range of normal commerce. Typically, either specialized government agencies or regulated monopolies are the owners and operators of telecommunications facilities, which provide, for a fee, message carrying services to the public at large; the public then uses those carrier facilities. Depending on the type of message, the use, in the past, was either telephone service or telegraph service.

This user concept, upon which the telecommunications industry has long relied, does not have much vitality any longer as far as many organizations are concerned. In the international marketplace, where an array of service choices have become available in the wake of advances in electronic technology, the transnational organizational user no longer is faced with the once rigid choice of either accepting or foregoing a particular service offered by a carrier at a fixed price. Because transnational organizations can and are deriving benefits of economy, efficiency, and control in a telecommunications marketplace of abundance and versatility, they are having a dynamic impact on the industry.

The most important recent trend is the reduction of the service role of carriers. By leasing circuits from carriers, an increasing number of major organizations have established their own service networks. Meanwhile, some entrepreneurs who also operate their own private networks using carrier provided circuits have begun selling services. Both the organization and the entrepreneur still are classified as users by the carriers, although what are used are only their circuit facilities, not their services. (The significance of this development will be discussed below.)

What has prompted this change has been the coercive fact of technology, which brought telecommunications into an era of abundance. The previous rationale—that carriers must provide end-to-end services in order to assure fair distribution of scarce benefits—has eroded with the recent advent of plentiful circuit facilities (Fig. 3–2). The public, of course, still relies on carriers for service, and indeed, the majority of carrier business remains in the public service sector. However, in the international sphere, where the collective business of carriers has been booming in recent years (see Fig. 3–1), a once wide gap between carrier service to the public and the leasing of circuits for private service has been narrowed considerably.

The growth rate of leased private circuits far exceeds that of the public system, as evidenced by the fact that in 1953 the ratio of reve-

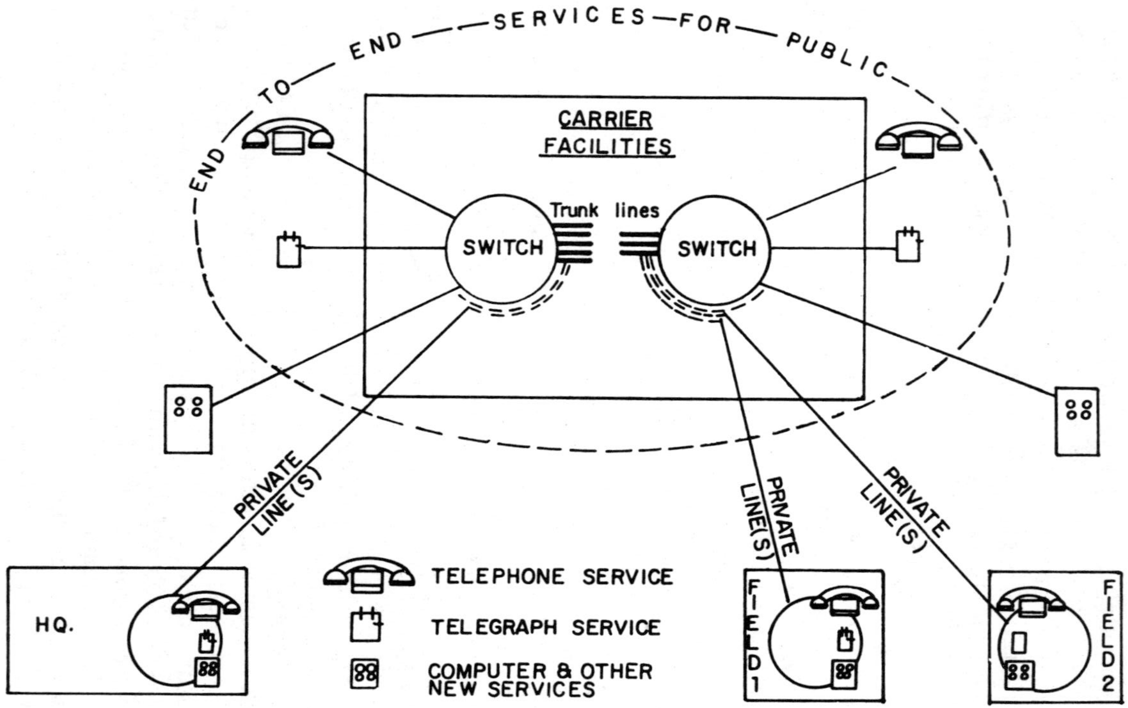

Figure 3–2. An End to Carrier Monopoly on End-to-End Services

Carriers have traditionally provided end-to-end telecommunications services and they still do provide telephone and telegraph services to the general public. Today, however, organizations can have their own service networks which use private lines leased from carriers or they can buy computer and other services from companies which also lease circuit facilities from carriers.

nues earned by U.S. carriers internationally from public service was four times that of private circuits, while two decades later the ratio had been halved. Put another way, during the early 1970s, as America's international telecommunications business passed the half-billion dollar mark, leased circuits were accounting for about one-third of total carrier revenues. Private service networks, in other words, are no longer playing an inconsequential role in international telecommunications. They are no longer exceptional or limited to specialized public interest activities, as are news agencies; they are becoming commonplace.

HOW MANY PRIVATE NETS, AND WHY

The number of private international telecommunications networks is not easily ascertainable and attempts at making a reasonable estimate are hampered by a reluctance of carriers to disclose relevant, albeit proprietary, market information. However, AT&T, at the request of the author, did undertake to determine the number and kinds of circuits it has between the United States and Canada. There are, according to the phone company, no fewer than 3,850 (of a total of 8,750) lines between the two countries given over for private service. These were broken down as 250 private-line-alternate-voice-data, 950 private-line-data, and 2,650 private-line-voice. How many cross-border networks exist remains a matter of conjecture, for obviously some of these circuits form the simplest kind of private network, which is point to point between, say, a corporate headquarters in New York and a small branch office in Toronto, while batches of other circuits are incorporated into worldwide networks such as the Pentagon's AUTOVON net.

From a policy perspective what is significant about these figures is that while AT&T has complete control over circuit *facilities* to Canada, only 56 percent of those facilities are devoted to the traditional public service network. It should be said that the phone company may be providing services as well as facilities under some leased circuit contracts, but since the FCC's precedent-setting "Carterfone" decision (1968), which lifted interconnection restrictions, the telephone company can gain control over nonpublic services only through success in the competitive marketplace and not as a regulatory right.

While the number of private networks crossing the 49th parallel may be unusually high, the phenomenon is far from uniquely U.S.–Canadian. Among the indicators that private nets are becoming an important worldwide dimension of international communications was a decision by Frost and Sullivan, a communications consulting firm, to publish a world telecommunications directory (at a price of $750 per copy) as an aid for managers of private nets. According to a Frost and Sullivan spokesman, a trend in recent years of establishing private nets reached a high point in the first half of 1975 when "a record number of overseas and international private line circuits" was installed by corporations and government agencies. But this development is being hampered, said the spokesman, by "the lack of information about foreign communication services and tariffs"—something for which the directory is being promoted as a remedy [11].

As new submarine cables and satellites have added both capacity and reach to the world's telecommunications facilities, the feasibility of building truly global private service networks exists. In practice, however, networks (a) are arranged to fit the activities of transnational users, and (b) their aggregate reveals that the mainstream in the flow of information on these networks is within the industrialized, western countries.

In the diagram of the UPI network (Fig. 3–3), note that UPI—one of the world's five general news agencies—has not extended its network to a black Africa, where it must rely on public telecommunications. And the diagram of Chase Manhattan's network (Fig. 3–4), while at first sight quite extensive, does not include any communist country, nor much of the Third World, nor in fact every western country. Moreover, the accompanying charts on submarine cables and pathways of communications satellites (Figs. 3–5 and 3–6) disclose that the trunk line capacity for international (as distinct from intracontinental) networks mainly stretch among the industrialized west. By comparison, there are few connections with the developing world and almost none with communist nations.

It is precisely because so much capacity became available that private networks began to proliferate. As one veteran of the carrier industry remarked: "We use to sell service, then speed, now bandwidth" [12] (a technical term for circuit capacity). The UPI diagram illustrates what an organization may do when it leases bandwidth. The news agency's leased voice-grade circuits can carry audio news reports, or telephoto transmissions, or multiple teletypewriter channels or combinations. Typically an organization which has leased alternate-voice-data circuits uses its network during the day to provide telephone service among its employees and at night to transmit sales orders, payroll information, and other types of data traffic among its various offices.

In building their networks, the means of transmission, save for the basic distinction between the spoken and written word, tend to be inconsequential—indeed arcane—to all but the organizations' communications managers. This level of consciousness creates an environment in which transnationals develop internal systems that bypass the constraints of the artificial divisions within the communications carrier industry. This is not to suggest that development is orderly; the contrary may in fact be the rule. Typically, several network operators interviewed for this study [13] said that their networks grew in a haphazard way, and that overall their nets were hybrids of voice, teletypewriter, and computer communication which have been acquired and integrated into a service system in response to newly perceived management information requirements, available technologies, and in pursuit of cost efficiencies. The resulting networks, especially the biggest (which also may have been the most cost efficient), do not remain static. Conversations at a major oil company with a $50 million a year communications budget; at a smaller, but still major, multinational firm which has a $17 million budget; and at a fledgling multinational corporation which spends $7 to 8 million annually on communications conveyed these impressions.

Given the prices of leased circuits (a transatlantic alternate-voice-data circuit runs about $10,000 a month), volume of traffic is usually an important consideration in deciding whether to operate a private service network. But where volume is high, this now available user option may be very cost effective. In the United States, for instance, one airline estimates that it saves about $400,000 a month on telecommunications costs through its participation in a network cooperative [14]. But beyond costs there are difficult to measure (but nonetheless significant) incentives to operate a private network: efficiency and control. While the value of either escapes being precisely quantifiable, the evidence, although anecdotal in nature, is persuasive to the effect that private networks have become the very lifelines of many transnational organizations.

It is through the network that much of the collective knowledge of an organization is shared internally; the network enables executive officers to acquire relevant decision making information, to effectively communicate their decisions, and receive feedback to them; the network may become so vital that loss of contact with an element of the organization forebodes misfortune (broken communications between an airliner and ground control, between a forward firebase and higher headquarters, between a money trader and the exchange). Even a degraded but partially functioning service network portends trouble as when messages are not discretely routed or certain types of communications cannot be transmitted. An organization which sells information, such as a news agency, values the control and efficiency of a private network because it enables the

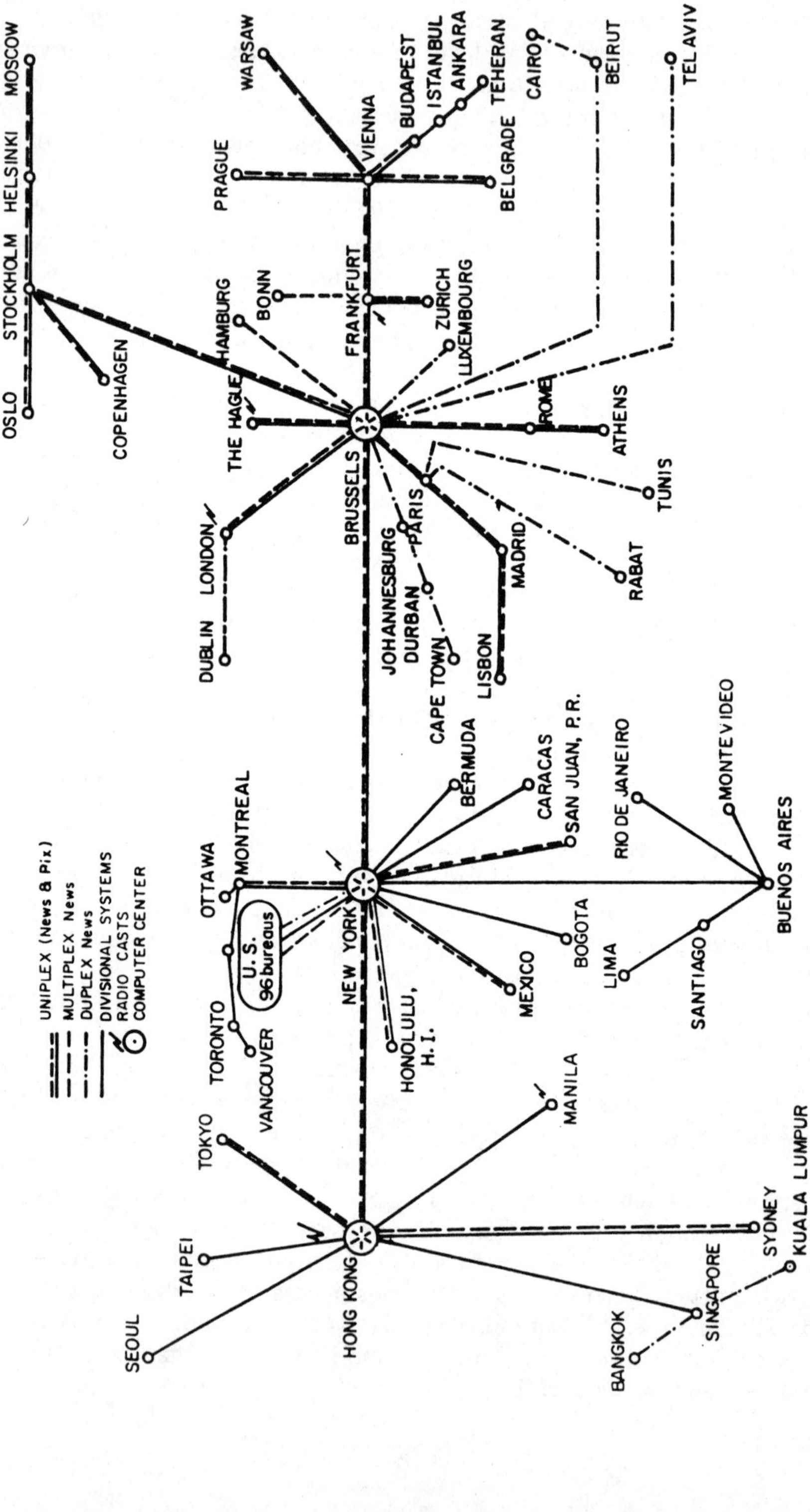
UNIPLEX (News & Pix)
MULTIPLEX News
DUPLEX News
DIVISIONAL SYSTEMS
RADIO CASTS
COMPUTER CENTER
OSLO
STOCKHOLM
HELSINKI
MOSCOW
COPENHAGEN
DUBLIN
LONDON
THE HAGUE
HAMBURG
BONN
PRAGUE
WARSAW
BRUSSELS
FRANKFURT
VIENNA
JOHANNESBURG
DURBAN
CAPE TOWN
PARIS
ZURICH
LUXEMBOURG
BUDAPEST
ISTANBUL
ANKARA
LISBON
MADRID
ROME
BELGRADE
TEHERAN
CAIRO
RABAT
TUNIS
ATHENS
BEIRUT
TEL AVIV
MONTREAL
OTTAWA
TORONTO
VANCOUVER
U.S. 96 bureaus
NEW YORK
HONOLULU, H.I.
MEXICO
BOGOTA
BERMUDA
CARACAS
SAN JUAN, P.R.
LIMA
SANTIAGO
BUENOS AIRES
RIO DE JANEIRO
MONTEVIDEO
TOKYO
TAIPEI
SEOUL
HONG KONG
MANILA
BANGKOK
SINGAPORE
SYDNEY
KUALA LUMPUR

Figure 3–3. United Press International's International Circuits

UPI's international private line network includes switching centers in Brussels, Hong Kong and New York. A national private line network also fans out of New York, where computers do editing and filing as well as switching. The diffusion of electronics technology since World War II has given heavy telecommunications users, like UPI, additional means for reducing their costs and also for improving efficiency and assuring control over the information flow. By installing at their own switching centers so-called multiplexing equipment readily available from many sources, they may configure private voice lines to provide telephoto and teletypewriter as well as audio (for radio broadcast) services.

Source: United Press International.

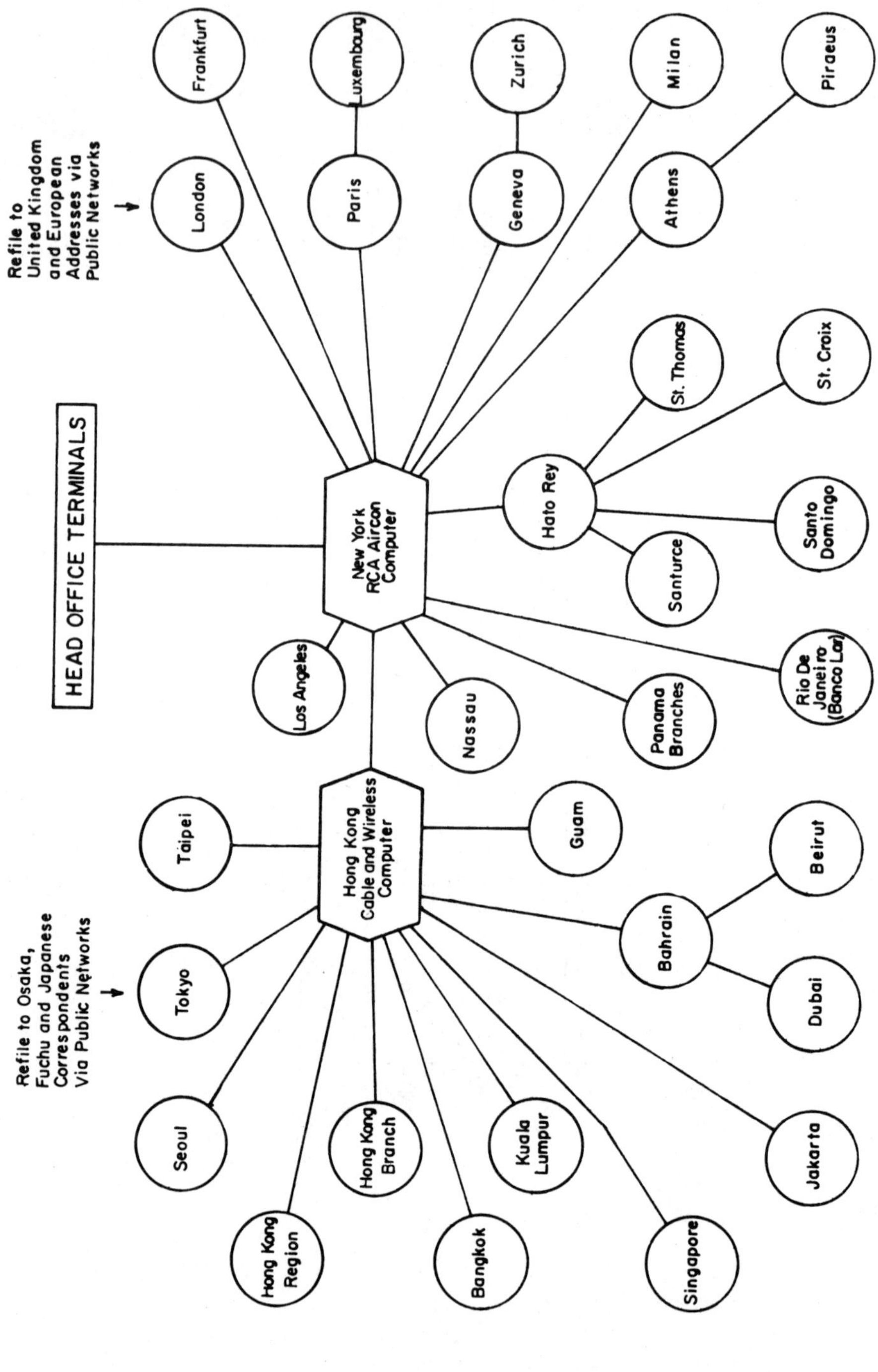
HEAD OFFICE TERMINALS
New York RCA Aircon Computer
Hong Kong Cable and Wireless Computer
Refile to United Kingdom and European Addresses via Public Networks
Refile to Osaka, Fuchu and Japanese Correspondents Via Public Networks
London
Frankfurt
Paris
Luxembourg
Geneva
Zurich
Milan
Athens
Piraeus
St. Thomas
St. Croix
Hato Rey
Santo Domingo
Santurce
Rio De Janeiro (Banco Lar)
Panama Branches
Nassau
Los Angeles
Taipei
Guam
Tokyo
Seoul
Beirut
Bahrain
Dubai
Jakarta
Hong Kong Region
Hong Kong Branch
Bangkok
Kuala Lumpur
Singapore

Figure 3–4. Chase Manhattan's International Network

Chase Manhattan established this network for teletypewriter traffic between the bank's New York headquarters and its foreign branches. Unlike UPI, which has its own computers, Chase has contracted for computer switching services from two carriers, RCA and Cable and Wireless. Refiling of messages is necessary in two instances: in Japan because of technical incompatibilities; in Britain because of a restrictive policy by the telecommunications authority there.

Source: Chase Manhattan.

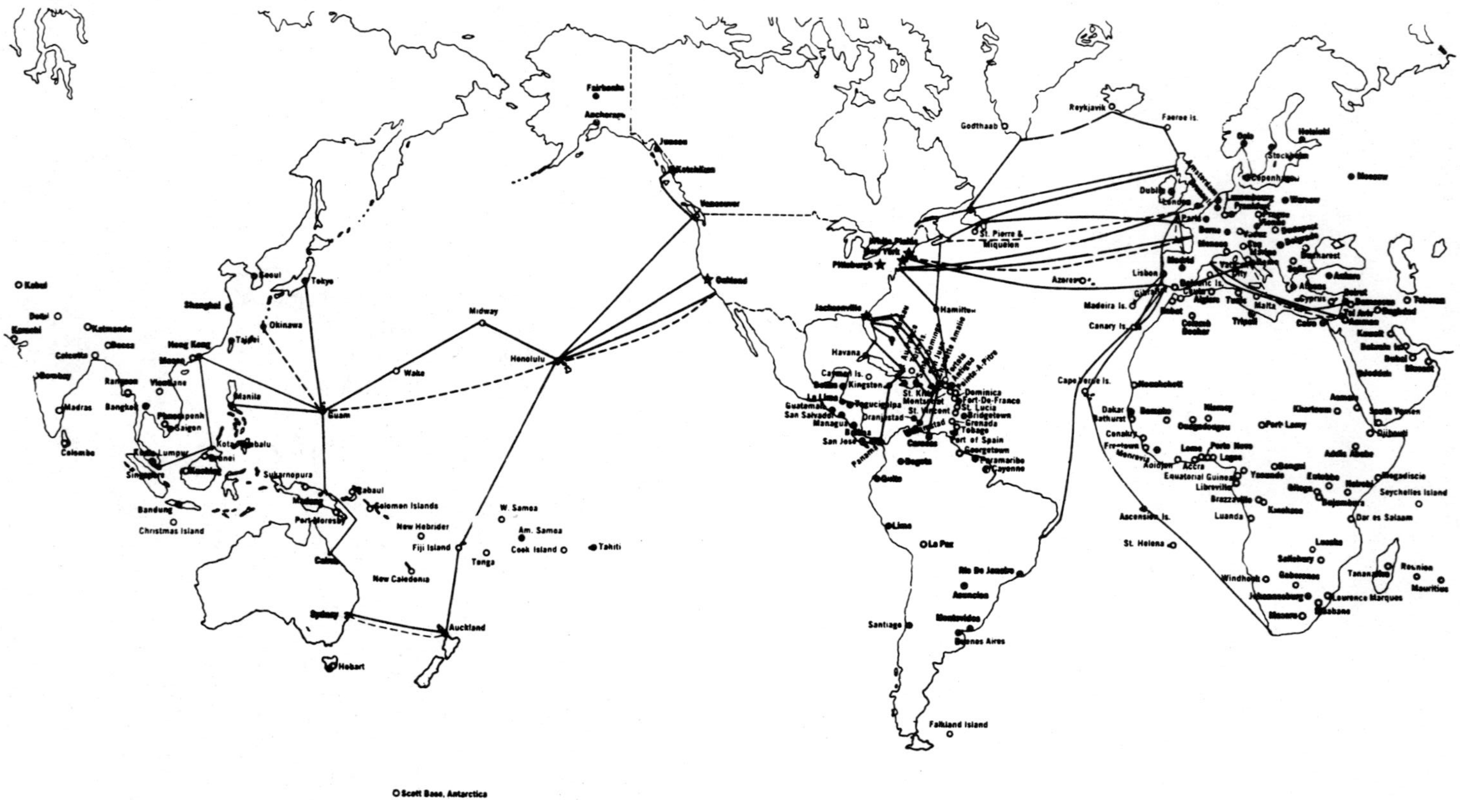

Fairbanks
Anchorage
Juneau
Vancouver
Oakland
Honolulu
Midway
Wake
Guam
Tokyo
Seoul
Okinawa
Taipei
Shanghai
Hong Kong
Manila
Kabul
Karachi
Calcutta
Bombay
Madras
Colombo
Rangoon
Bangkok
Vientiane
Phnompenh
Saigon
Kuala Lumpur
Singapore
Brunei
Bandung
Christmas Island
Sukarnopura
Rabaul
Solomon Islands
Port Moresby
New Hebrides
Fiji Island
New Caledonia
W. Samoa
Am. Samoa
Tonga
Cook Island
Tahiti
Sydney
Auckland
Hobart
Scott Base, Antarctica
Godthaab
Reykjavik
Faeroe Is.
St. Pierre &
Miquelon
New York
Pittsburgh
Hamilton
Azores
Jacksonville
Havana
Cayman Is.
Kingston
Belize
La Lima
Tegucigalpa
Guatemala
San Salvador
Managua
San Jose
Panama
Dominica
Fort-De-France
St. Lucia
St. Vincent
Bridgetown
Grenada
Tobago
Port of Spain
Georgetown
Paramaribo
Cayenne
Oranjestad
Caracas
Bogota
Quito
Lima
La Paz
Rio De Janeiro
Asuncion
Montevideo
Buenos Aires
Santiago
Falkland Island
Amsterdam
Dublin
London
Paris
Lisbon
Madrid
Oslo
Stockholm
Helsinki
Copenhagen
Moscow
Warsaw
Prague
Budapest
Belgrade
Bucharest
Sofia
Athens
Cyprus
Beirut
Damascus
Tel Aviv
Amman
Baghdad
Tehran
Cairo
Tunis
Malta
Tripoli
Algiers
Madeira Is.
Canary Is.
Cape Verde Is.
Dakar
Bathurst
Conakry
Freetown
Monrovia
Abidjan
Accra
Lome
Porto Novo
Lagos
Niamey
Ouagadougou
Bamako
Nouakchott
Fort Lamy
Khartoum
Asmara
South Yemen
Djibouti
Addis Ababa
Mogadiscio
Nairobi
Entebbe
Kigali
Bujumbura
Seychelles Island
Dar es Salaam
Equatorial Guinea
Libreville
Brazzaville
Kinshasa
Luanda
Ascension Is.
St. Helena
Lusaka
Salisbury
Gaberones
Windhoek
Johannesburg
Maseru
Laurence Marques
Mbabane
Tananarive
Reunion
Mauritius
Kuwait
Bahrain Is.
Dubai
Muscat

Key

Cables: ———— Existing

- - - - - - - - - Proposed

Figure 3–5. Submarine Telephone Cables

Undersea cables have long been used for telegraph service and even as early as 1921 there was telephone service by cable between Florida and Cuba. But the modern period of submarine cable began with transatlantic telephone service between the U.S. and Europe by cable in 1956. For the U.S., undersea cables are links to the richer countries of the world and to its interests in the Caribbean. U.S. telephone service to other parts of the world is by direct radiophone or satellite (as indicated by "0") or via an intermediate point (as indicated by "0").

Source: AT&T Long Lines, October 1973.

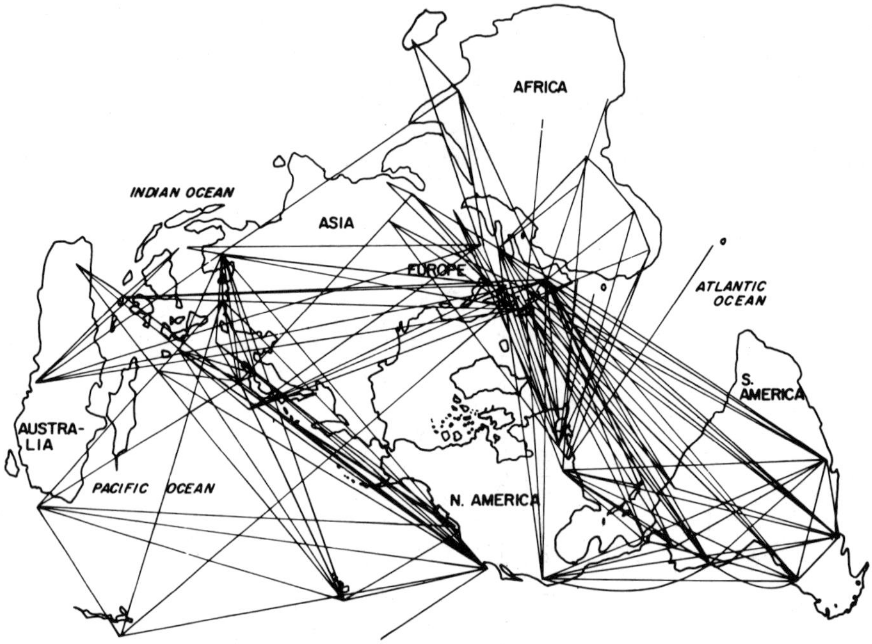

Figure 3–6. The Global Communications Satellite System

The growth of the international satellite system from a single North Atlantic pathway in June 1965 to 325 pathways covering much of the world by June 1974 has continued to reflect greater communications density among the industrialized western countries.

Source: COMSAT, 1974 Report to the President and the Congress.

organization to insure proper distribution of its various types of information.

The intangible but real benefits of control and efficiency are further evidenced by the presence of communications leagues such as the Société Internationale de Telecommunications Aeronautiques, which serves the airlines of the world. That leagues are formed not merely to cut costs for their communications dependent members may be seen in the diagram of the under-construction SWIFT network for the international banking community (Fig. 3–7). Note that not only are there seven "backup connections" but also two switching centers with twin primary connections. Beyond the added costs of redundancy, the member banks of SWIFT will also pay a substantial premium for building into their system the capability of transmitting encrypted messages.

So the point at which a league is formed or a private network becomes desirable is not merely where a rising demand curve for information intersects with the still descending curve of transmission costs—the less calculable factors of control and efficiency benefits are inputs as well. And the value of each factor may change over time.

At one multinational corporation, for instance, the comptroller said his company headquarters receives some 250 financial line item reports periodically from its numerous plants and offices. The possibility of such detailed reporting, he explained, arose when computer communications made it economically feasible. But today that information flow, made possible by electronic technology at an acceptable cost, is considered, according to a front office executive, "critical . . . imperative . . . vital" to the corporation's well-being [15].

SERVICE BUREAUS

It is not necessary, however, for an organization which wants to make use of various computer services to either operate its own network, participate in a league, or install a computer at every point of actual use. The same electronic technology which led to an abundance of telecommunications circuits also gave birth to a new service commonly called teleprocessing.

Non-carrier entrepreneurs who recognized that computer services could be made available at a distance by linking computers to telecommunications circuits sell remote processing services which their customers access through carrier facilities.

The largest of these so-called service bureaus is generally acknowledged to be General Electric which operates by its self-description a

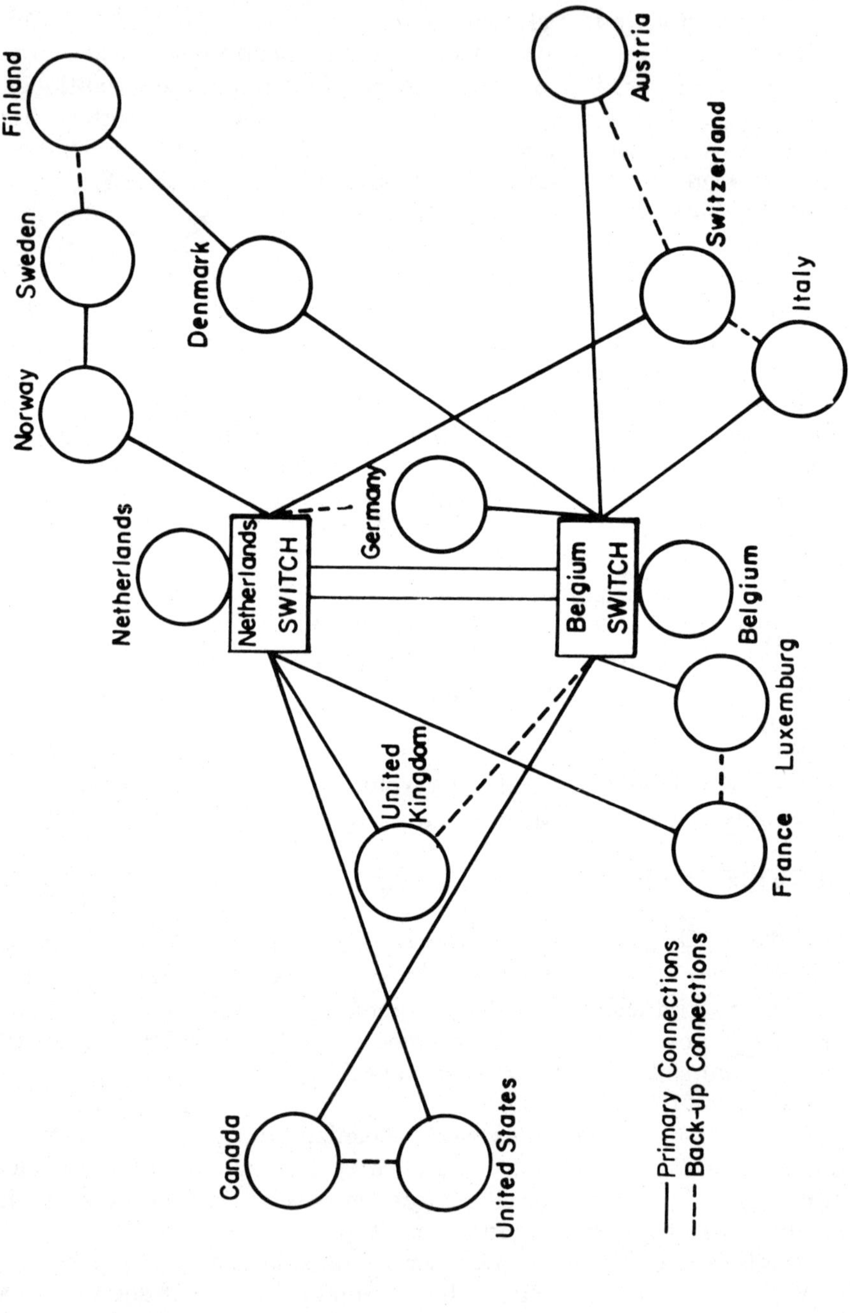
Finland
Sweden
Norway
Denmark
Austria
Switzerland
Italy
Netherlands
Netherlands SWITCH
Germany
Belgium SWITCH
Belgium
Luxemburg
United Kingdom
France
Canada
United States
—— Primary Connections
- - - Back-up Connections

Key

——— Primary Connections

- - - - - Back-up Connections

Figure 3–7. The SWIFT Network

On May 3, 1973, the Society for Worldwide Financial Telecommunications (SWIFT) was organized under Belgian law as a nonprofit cooperative of over 250 banks in Europe and North America for the purpose of improving the transmittal of international payments messages. These had been handled mainly by mail and to a lesser extent by cable. The SWIFT network, scheduled to become operational during 1976, will use computer/communications in order to transmit more efficiently an estimated 300,000 messages per day by 1978.

Source: The Magazine of BANK ADMINISTRATION, September 1975, p. 37.

"billion-dollar system [which can be accessed] with local [telephone] calls in over 400 cities in North America, Europe, Japan, and Australia" [16].

In the same category as GE is Control Data Corporation, which signed a private consent decree with IBM that took the world's leading computer firm temporarily out of the U.S. domestic service market. IBM's World Trade subsidiaries do, however, offer similar services outside the United States, in Canada and Europe. Some other American service bureaus also are active internationally. But, according to a recent industry report [17], they should be considered "second level" providers of "network information services."

Network information services—i.e., the services provided commercially by vendors whose business is enabling multiple customers to realize cost efficiencies by linking them to a shared computer facility with telecommunications circuits—are a modern phenomenon. The first practical computer time sharing system is said to have been demonstrated in 1964 at Dartmouth, with commercial application the following year. GE points to itself as a pioneer, having built a 42-city U.S. network in 1969 which was extended to Britain in the next year. By 1975, the GE net was worldwide with more than 5,000 customers in eighteen time zones. According to the computer service industry annual report of that year, such services in 1973 generated a total of $766 million in revenues, with a very strong growth rate (by the end of the decade total revenues were forecast to quadruple). Information made available by a separate source indicated that the ratio of U.S. domestic to foreign revenues was roughly five to one. However, all projected growth rates abroad were higher than in the United States.[a]

"A major underpinning of the growth in computer services," the industry report said, "is the change in attitude on the part of many users . . . to spend more effort scrutinizing their EDP (electronic data processing) costs." This tendency, said to have been growing since 1968, has created a healthy atmosphere for computer utilities. The utility presents itself to potential customers as a less expensive alternative to having one's own computer facilities or going without. The utilities can do this even though the necessary telecommunications channels that link customers to central computer facilities are costly (11 percent of total vendor expenses is cited as typical) [18]. The utility, however, more than offsets the telecommunications costs by

[a]There is a suggestion here for further research by analogy—i.e., it might be useful to identify other industries which experienced a similar pattern of development in order to gain some insights into how network information services might mature internationally.

(a) acquiring leased circuits at what are, in effect, discounted rates, and (b) skillfully spreading out the work load at the computer facility by such strategies as marketing in many time zones and offering price reductions for off-peak processing.

For illustrations of the service bureau or computer utility concept, the author requested and GE provided some sample application descriptions used by GE in promoting their service. Five synopses of the GE-supplied material are offered below.

1. For Kelsey-Hayes, a Windsor, Ontario manufacturer of steel wheel assemblies, the daily output of inspection and quality control reports totaled 1,100, a mountain of paperwork which not only was labor consuming to handle but gave rise to costly waiting times when crucial information was needed. One option for the Canadian firm was to buy its own computer in order to process its reports more efficiently. An alternative was time sharing, a scheme whereby Kelsey-Hayes could both avoid the initial capital expenditure for a computer and also pay only for the amount of computing time it actually used. GE, with a computer "super center" in Cleveland, Ohio, and a dedicated telecommunications circuit within the plant's local telephone dialing area, was the option chosen by the Canadian firm.

2. One role of Merrill Lynch Royal Securities of Canada is that of a "jobber" in the money market where corporations with idle or excess funds, for instance, make short term loans to borrowers. It is a business with a fine margin between profits and loss for investment dealers such as Merrill Lynch. Until a few years ago, the firm kept track of due dates, calculated rates of return, and tallied its profit/loss position in accordance with changing bank rates manually. Now it uses GE. One person instead of three are needed. Not only does a computer in Cleveland prepare the next day's trading blotter, but it does it after 9 pm at a discounted rate. For the jobber, however, what is more important is that he knows whether he made money the day before or not. The old manual system left that question unanswered for weeks.

3. In 1969 the First Wisconsin National Bank decided to open a foreign branch in London. Unlike large multinational banking firms, a private international network for financial consolidation and reporting was prohibitively expensive. For the Milwaukee bank even a daily telex message had to be kept brief because of transmission rates. When foreign currencies began floating, First Wisconsin felt it had to have an improved reporting system. GE provided the answer, since both the London branch and the Milwaukee headquarters

could build and access a common data base at GE's Cleveland "super center."

4. Star Shipping of Norway had managed its worldwide business through the company's noncomputerized communications system, which fed information from Star offices in North America, Europe, and Japan to its headquarters in Bergen. But a growth in business was overwhelming the system. To get better control of its scheduling—which for Star meant giving branch offices more autonomy and authority while retaining the headquarter's ability to do overall monitoring—the international shipping company contracted for global network information service from GE.

5. Even the biggest of private multinational enterprises may turn to a computer utility rather than further expand their own networks to accommodate a desired cross-border data capability. Mobil Oil has done that. The major petroleum company uses the GE system for consolidation and processing of financial data from affiliated companies around the world in order to better prepare short range profitability reports for management.

Another example of the use of computer/communication services is the case of four East European airlines. Balkan Bulgarian Airlines, Czechoslovak Airlines, Malev Hungarian Airlines, and LOT Polish Airlines all use Control Data Corporation's time sharing computer facility in Atlanta for reservations, although one wonders whether those airlines are even aware of that fact, since the service is provided through a third party. The airlines actually contract with SITA, the international communications league for air carriers, to provide the service. SITA, however, subcontracts with CDC. The SITA communications network can put, say, a LOT reservations clerk in Warsaw on-line with CDC's Atlanta computer within a few seconds without the clerk (or, for that matter, anyone else unless they inquired) knowing where the computer was physically located. SITA also provides this service for Icelandic and VARIG airlines.

NEW POLICY QUESTIONS

The changing conditions of international telecommunications bring forth a pair of policy questions, which may be separated into categories of high and low politics. At the upper level are questions about the nexus of telecommunications, the conduct of transnational organizations, and the impact of those organizations on international affairs. At the other level is the more particular question of how telecommunications administrations (as carriers are known internationally) are being affected.

High Politics

What we know about high politics begins with an appreciation of global interdependence, the role of transnational organizations in bringing that condition about, and the reliance those organizations have on telecommunications systems. Thereafter, unfortunately, less is known. Yet, an organization wishing to strengthen its transnational posture may very well have to factor politics into its telecommunications decision making.

For illustrations in the nongovernment sector we can look at cases of two multinational corporations in which a decision for both was whether to merge their U.S. and Canadian computer/communications networks [19]. Company A owns a sizeable Canadian subsidiary which has a history of being somewhat autonomous from the parent corporation. The subsidiary has its own remote job entry network controlled by a computer center in Montreal. When the parent proposed an integration plan, the subsidiary balked. The reason cited was the attitude of the Canadian government, which so far has not actually banned such integration, but has hinted that it might adopt such a policy. The parent company, although it suspects the subsidiary's true reason for balking is an attempt to retain as much independence as possible, has continued to move cautiously because of the murky political waters north of the border. The result is that planning for integration has proceeded without an actual decision to tie the subsidiary's network into a U.S. computer center.

Company B, on the other hand, has overcome similar misgivings in the face of what it considered a compelling economic gain. Like the other multinational, Company B had a computer center in Canada but it was not originally established for the same reason. In 1972, B, which does not have a subsidiary with a tradition of autonomy, established a center in Toronto because the company sensed, according to its assistant communications manager, that "Canada for Canadians" was the prevailing attitude regarding computers. Indeed, a year later the Ottawa government published a position paper that read in part: "The Government attaches great importance to the development and support of a strong data processing industry in Canada and will favor measures to maximize Canadian influence in key areas of computer/communications" [20].

But in 1975, only three years after establishing a computer center in Toronto, and in the face of Ottawa's published position, the center was closed and replaced by a terminal attached to a dedicated private line which terminates at a U.S. computer center. Asked to explain, a company official said that the separate center in Toronto was economically inconsistent with a corporate decision to centralize its computer programs. The result is that inventory, payroll, financial, and personnel files about the multinational's Canadian operations that once were processed in Toronto, now are done within the United States—because it is cheaper.

The political factor is not peculiar to Canada, as evidenced by the fact that Hong Kong and Brussels have emerged as international telecommunications hubs as transnational communicators moved to frustrate more restrictive national telecommunications policies elsewhere in Asia and Europe. "User evasion" (the phrase of a British Post Office official) is boosting Brussels toward becoming the capital of Europe and giving colonial Hong Kong a more secure position as a tennant on territory leased from China.

Low Politics

At the traditional political level of international telecommunications the impact of change is, as might be expected, felt more fully. The never easy planning of costly new facilities has been complicated for carriers due to their looser control over end-to-end services. Meanwhile, what some carriers perceive as outsider infringement on their sources of revenue has bred tensions.

Upgrading and building new facilities is a major task approached by carriers with considerable care since the costs are extremely high (in terms of assets, AT&T is the largest U.S. corporation). In order to do facility planning, carriers engage in the somewhat arcane business of trying to reasonably forecast traffic. While a number of statistical methods are used, the essential goal is to predict service demand far enough ahead so that there is sufficient lead time to build or upgrade the facilities required to meet the anticipated demand. Those who labor in these fields rely on such information as minutes of usage of circuits, message counts, number of terminals connected to circuits, and the revenues they produce. All this data, of course, relates to service.

But when transnational organizations are providing their own services—or at least are not acquiring them from carriers—the traditional forecasting techniques are thwarted. When all a user wants from a carrier is a circuit, then, as a recent FCC Working Group on Forecasting concluded, "It is difficult to project the number of private line voice-grade circuits needed in the future through the utilization of sophisticated statistical and econometric techniques" [21].

When carriers don't provide service, then "forecasting is a matter of best marketing judgment," according to the Working Group. Curiously, this government group—in which the carriers heavily participated—found that even the U.S. government's own large requirements for international voice/data circuits was "impossible to accurately project. . . ."

For carriers the dilemma is this: on the one hand their loss of control over service compounds their difficulty in making reasonable traffic forecasts in order to do essential facility planning, while on the other hand their abundant supply of circuits leads to more leasing and, in turn, less control. In this connection, data communication is particularly perplexing. In the computer era machines can communicate with each other via telecommunications circuits at enormously varying speeds ranging from 50 bits per second (slow teletypewriter) to 9,600 bits per second and higher (fast computer). But by far the most widely available circuit facilities are telephone lines, which have bandwidths that are routinely capable of handling 2,400 bits per second transmissions.

Since carriers tend to feel that data communications has a very strong future in telecommunications, they naturally want to prepare themselves to tap that seemingly lucrative market. To do so they must attempt to ascertain how many circuits will be needed, at which bandwidths, and between or among what locations. Pondering such questions is extremely difficult because data services have developed apart from the carrier industry. Nonetheless, carriers are taking forward steps, such as Western Union International's International Digital Data Service. But the uncertainties are enormous. AT&T, for instance, after considerable effort was unable to quantify the extent to which data traffic flows between the United States and Canada where, as we have seen, the number of private circuits is not much less than those available to the public. Indeed, AT&T officials noted that a paucity of reliable information exists about the number and use of data transmission devices.

This information void points also to another apprehension (felt perhaps more strongly abroad than in the United States) about losing service which the carriers feel rightfully belongs to them. Carriers tend to be especially leary about computer service bureaus lest they engage in merely message switching and thereby erode carrier telex revenues. In a celebrated case over this issue, Tymshare was reprimanded by telecommunications authorities in Britain. Other protectionist steps taken by carriers have included restrictions on the type and number of terminals which can be connected to a private international circuit and discriminatory tariffs.

ACTORS	PERSPECTIVES	SUBSTANTIAL CONCERNS	INTERNATIONAL FORUM
Governments	National	Improved service for constituents; Sovereignty; Trade; Influence; Socio-economic impacts	ITU OECD
Carriers	National	Prices and practices for interconnection of voice and record services; Control over services	ITU CEPT (West Europe Regional)
Manufacturers	National; some transnational; New electronic, emphasizing transnational	Trade	ITU

Transnational Organizations	Transnational	Cost Cutting; Efficiency; Control	NONE
International Service Bureaus	Transnational	Trade with nondiscriminatory universal market access.	ITU

The shaded areas indicate additions since the mid 1960s.

Figure 3–8. Policy Making in Two Eras of International Telecommunications

When telecommunications around the world were scarce and services few, policy makers shared the same forum and each championed his national interests. Today, however, when telecommunications are abundant and services versatile, nationalism confronts transnationalism. Two additional forums have been added—OECD and CEPT—and significant new actors have appeared—international service bureaus and transnational organizations, with the latter as yet participating in neither the old nor the new forums.

When they can, the managers of private international networks attempt to evade these policies, for example, by reconfiguring their networks.

CONCLUSION

The overall result is a curious international game of high and low politics with which, it may be said, neither the transnational organizations nor the national carriers are very pleased.

Take, for example, the case of SWIFT. Briefly stated, the European and American banking communities, central players in the reordered game of international economic order, decided several years ago that they needed their own international computerized telecommunications system to handle funds transfer. So they formed the Society for Worldwide Interbank Financial Telecommunications (SWIFT) for the purpose of establishing a network similar to the U.S. Bank-Wire system. The concept sent shock waves through the telecommunications establishment.

According to several accounts, both the European Post, Telegraph and Telephone administrations (PTTs) and the private U.S. telegraph or record carriers initially took the position that unless they were in charge of the network, and thereby could not only prevent a loss of some lucrative telex revenues but also acquire added business, there would be no SWIFT. They miscalculated, for bankers have more political clout than carriers. Still, the carriers were able to wrestle a complex—some say contorted—arrangement with the key provision being that after five years the telecommunications administrations can, in effect, take over the bank system, provided that the carriers are able to maintain the same standards of service.

The deal, signed in May 1975, did not quell all the dissidents, especially the French PTT, which showed an intention to challenge every step taken toward implementation of the network. This attitude certainly must have caused some embarrassment for SWIFT's president because he also heads the Banque Nationale de Paris. The assumption in the banking community was that the French economics ministry, formerly headed by President Giscard D'Estaing, would prevail within the upper councils of the French government on behalf of the Banque. If not, France would either suffer the consequences of being an outsider or have to interconnect to SWIFT through a third country. Obviously, neither alternative would be attractive to Banque Nationale de Paris. But then, neither is SWIFT attractive to the French PTT, which is said to cross-subsidize its deficit ridden postal system with profits from PTT-controlled international telecommunications.

The SWIFT case broadly illustrates how problematic international telecommunications policy making can be. In the wake of recent world monetary troubles, the high politics of that situation would seem to favor a technological fix, i.e., creation of the network. But the low politics of implementation soon became embroiled over the issue of who would control the network. Bankers began by insisting on control of their service, the carriers deemed mere leasing of their circuits insufficient. Eventually the parties agreed to the possibility that the balance might be shifted some day from the former to the latter.

But will it? If the currently fashionable notion that information is a resource takes hold, will bankers, or anyone else with high stakes in the viability of their information resources, be willing to return to traditional communications concepts and modes?

This conceptual adjustment suggests the necessity for reorienting our thinking about how international telecommunications policy is made. No longer, it seems, can policymaking be the exclusive purview of carriers who have gathered periodically under the auspices of the International Telecommunications Union to consider and then make recommendations essentially related to the technical aspects of cross border information flows by radio or wire. Today technicians are but a single, albeit still important, set of policyplayers in a game that by many accounts is described as a "mess."

Policymaking has slid into a gray world in which uncertainties abound. While the various parties interviewed for this study viewed the situation with different biases, preferences and perspectives, all tended to agree that the institutions and organizations which have functioned in the past to guide and manage international telecommunications are proving inadequate in an era of abundance and versatility.

The basic reason is that the technology which has produced those conditions has enabled transnational organizations—who have been and still are apart from the policy process—to provide much of their own telecommunications service. Hence, the crucial question of control. The need is to look beyond the quantitative factor of abundant telecommunications facilities to the qualitative issues of who controls the systems, and for what benefits which are distributed to whom?

Notes

NOTES TO PART ONE

1. Additional details in: *Scientific American*, Vol. 227, No. 3, September 1972.

2. Edwin B. Parker and Marc Porat, *Social Implications of Computer/Telecommunications Systems*, Report No. 16, Center for Interdisciplinary Research, Stanford University, February 1975, Figure 2, p. 11.

3. Additional details in: William Read, *Foreign Policy: The High and Low Politics of Telecommunications*, Program on Information Technologies and Public Policy, Harvard University, Cambridge, Ma., February 1976, P–76–3.

Program on Information Technologies and Public Policy, *Information Resources: Foreign Policy Choices for the 1970's and 1980's*, Harvard University, Cambridge, Ma., May 1976, working paper.

4. Thomas Reed Powell, as quoted in: Jerold S. Auerbach, *Unequal Justice: Lawyers and Social Change in Modern America*, Oxford University Press, New York, N.Y., 1976, p. 9.

5. William Read, *Foreign Policy*.

Program on Information Technologies and Public Policy, *Information Resources*.

6. Theodore P. Seto, "The Early Years of the Telephone Industry: A Case Study in Technological Change," unpublished paper, Harvard University, January 28, 1972, p. 21.

7. Additional details in: Erik Barnouw, *A History of Broadcasting in the United States*, 3 vols., Oxford University Press, New York, N.Y., 1966–70.

Kurt Borchardt, *Structure and Performance of the U.S. Communications Industry: Government Regulation and Company Planning*, Harvard University, Boston, Ma., 1970.

Erwin G. Krasnow and Lawrence D. Longley, *The Politics of Broadcast Regulation*, St. Martin's Press, New York, N.Y., 1973.

8. Erik Barnouw, *A History of Broadcasting in the United States: Vol. 2, The Golden Web, 1933–53*, p. 33.

Additional details in: Vincent Mosco, *The Regulation of Broadcasting in the United States: A Comparative Analysis*, Program on Information Technologies and Public Policy, Harvard University, Cambridge, Ma., August 1975, P–75–76.

Vincent Mosco, *Reforming Regulation: The FCC and Innovations in the Broadcasting Market*, Program on Information Technologies and Public Policy, Harvard University, Cambridge, Ma., February 1976, pp. 76–82.

Erwin G. Krasnow and Lawrence D. Longley, *The Politics of Broadcast Regulation.*

9. In 1974 locally produced and originated programming, excluding commercials, accounted for a little more than 10 percent of the total operating hours of the average local station. Comments of the Association of Maximum Service Telecasters, Inc., pursuant to the *Second Notice of Inquiry*, Docket No. 20271, November 14, 1975. At the same time, revenues from *local* advertisers (as opposed to network ad revenues and national and regional spot time revenues) accounted for more than 23 percent of television gross advertising revenues. 45 *Television Factbook*, Services Volume, 1976, pp. 63–a.

10. *Clear Channel Broadcasting in the Standard Broadcast Band*, FCC Docket No. 20642, *Notice of Inquiry and Notice of Proposed Rule making*, released December 12, 1975, ___ FCC2d ____ (1975), Order No. FCC 75–1331.

11. *Notice of Inquiry and Notice of Proposed Rulemaking*, Docket No. 20642, ____ FCC 2d at ____ (¶18).

12. It was not until 1973 that the FCC was able to say that standard (AM) and FM broadcast services were to be treated together as a single aural service. See *Notice of Inquiry and Notice of Proposed Rulemaking*, Docket No. 20643, ____ FCC 2d ____, ____ (¶15, 1975).

13. Additional details in: William T. Knox, "Cable Television," *Scientific American*, Vol. 225, No. 4, October 1971, pp. 22–29.

14. Erik Barnouw, *A History of Broadcasting in the United States: Vol. 1, A Tower of Babel: to 1933*, pp. 143–144.

15. Sig Mickelson, "Technological Censorship," *Intermedia*, Vol. 2, No. 2, 1974, pp. 4–5.

16. Harold Laswell, *Politics—Who Gets What, When, How*, McGraw-Hill, New York, N.Y., 1936.

17. Adam Smith, *The Wealth of Nations*, Random House, New York, N.Y., 1937, p. 423.

18. U.S. Congress. Senate. Committee on the Judiciary, Subcommittee on Antitrust and Monopoly. Hearings on S. 1167, "Part 6: The Communications Industry," July 9, 30 and 31, 1974, p. 4201.

19. United System Service, Inc., *Response to First Supplemental Notice*, FCC Docket No. 20003, pp. 21–22.

20. Paul J. Garfield and Wallace F. Lovejoy, *Public Utility Economics*, Prentice-Hall, Englewood Cliffs, N.J., 1964, p. 45.

21. 47 U.S.C. §201, Common carriers, services and charges.

22. Eugene Rostow, personal communication, 1975.

23. 47 U.S.C. §151, Creation of the Federal Communications Commission.

24. Office of Telecommunications Policy, Executive Office of the President, "Regulatory Treatment of Communications Brokerage," Comments on FCC Docket No. 20097, February 24, 1975, p. 21.

25. *Ibid.*, p. 22.

26. *Smith* v. *Illinois Bell Telephone Company*, 282 U.S. 133 (1930).

27. United System Service, Inc., *Response to Notice*, FCC Docket No. 20003, April 1975, p. 5 and Exhibit 2.

28. 26 FCC 2d 247 (1970), Docket No. 18866, *Report and Order.*

29. Richard Gabel, *Development of Separations Principles in the Telephone Industry*, Institute of Public Utilities, Michigan State University, East Lansing, Mich., 1967, p. 147.

30. *Ibid.*

31. *Telecommunications Reports*, Vol. 41, No. 39, September 29, 1975, p. 30.

32. John R. Pierce, "Communication," *Scientific American*, Vol. 227, No. 3, September 1972, p. 37.

33. Additional details in: Vacroux, Andre G., "Microcomputers," *Scientific American*, Vol. 232, No. 5, May 1975, pp. 32–40.

34. Claude E. Shannon and Warren Weaver, *The Mathematical Theory of Communication*, University of Illinois Press, Urbana, Ill., 1949, p. 43.

35. Bell Laboratories and Western Electric, *Impact*, Bell Laboratories, Murray Hill, N.J., March 1971, p. 92.

36. Additional details in: Henri Busignies, "Communication Channels," *Scientific American*, Vol. 227, No. 3, September 1972.

37. Additional details in: Frank G. Heath, "Large-Scale Integration in Electronics," *Scientific American*, Vol. 222, No. 2, February 1970.

38. Anthony G. Oettinger, "Compunications in the National Decision-Making Process," *Computers, Communications, and the Public Interest*, Martin Greenberger (ed.), Johns Hopkins Press, Baltimore, Md., 1971, pp. 74–113.

39. 28 FCC 2d 267 (1971), *Decision and Order*, and 40 FCC 2d 293 (1973), *Order*, FCC Docket No. 16979.

40. American Newspaper Publishers Association, "Proposed Findings of Fact and Conclusions of Law," July 18, 1974, submitted on FCC Docket No. 19919.

41. *Associated Press* v. *United States*, 326 U.S. 1 (1945).

42. Stuart L. Mathison and Philip M. Walker, *Computers and Telecommunications: Issues in Public Policy*, Prentice-Hall, Inc., Englewood Cliffs, N.J., 1970, p. 104.

American Telephone and Telegraph Company FCC Tariffs 260 and 263.

43. 13 FCC 2d 420 (1968), Docket Nos. 16942 and 17073, *Decision.*

Stuart L. Mathison and Philip M. Walker, *Computers and Telecommunications*, pp. 85ff.

44. National Academy of Sciences, Computer Science and Engineering Board, *A Technical Analysis of the Common Carriers/Users Interconnections Area*, National Academy of Sciences, Washington, D.C., June 1970, pp. ii, 2.

45. Additional details in: W. Wesley Peterson, "Error-Correcting Codes," *Scientific American*, Vol. 206, No. 2, February 1962, pp. 96–108.

46. Stuart L. Mathison and Philip M. Walker, *Computers and Telecommunications*, p. 193.

47. *Ibid.*, p. 195.

48. *Ibid.*, p. 193.

49. Peter Shapiro, *Public Policy as a Determinant of Market Structure: The Case of the Specialized Communications Market*, Program on Information Technologies and Public Policy, Harvard University, Cambridge, Ma., September 3, 1974, W–74–10.

50. *U.S.* v. *Western Electric*, 1956 Trade Ca. 1, ¶68,246, pp. 71, 137.

51. Additional details in: 47 FCC 2d 644 (1974), FCC Docket No. 20097, *Notice of Inquiry*.

52. *Telecommunications Reports*, Vol. 41, No. 1, January 6, 1975, p. 1.

53. John T. Dunlop, "The Limits of Legal Compulsion," November 12, 1975, unpublished paper, pp. 3–4.

54. 2d Circuit, *Bell Telephone of Pa.* v. *FCC*, No. 73–1806, October 19, 1973.

55. Peter Shapiro, *Public Policy as a Determinant of Market Structure*, p. 69.

56. Order No. FCC 75–1043, Docket No. 19919, *Interim Decision and Memorandum Opinion and Order*, released September 18, 1975, p. 29, ____ FCC 2d ____ (1975).

57. *Ibid.*, p. 30.

58. *Ibid.*

59. *Telecommunications Reports*, Vol. 41, No. 39, September 29, 1975, p. 10.

60. Order No. FCC 76–30, Docket No. 19919, *Decision and Memorandum Opinion and Order*, released January 22, 1976, pp. 5–6, ____ FCC 2d ____ (1976).

61. *Ibid.*, p. 6.

62. American Telephone and Telegraph Company, Transmittal No. 12546, Tariff FCC Nos. 260, 264, and 266, filed April 19, 1976.

63. 44 FCC 2d 252 (1973), Docket No. 18684, *Order.*

64. The lineage of such arguments can be traced through FCC Docket Nos. 18128 (1969–   ), 16258 (1965–1971) and 14650 (1962–1967), among others.

65. FCC Docket No. 18128, *Recommended Decision of the Chief, Common Carrier Bureau*, released January 19, 1976, ____ FCC 2d ____ (1976).

66. *Ibid.*, p. 65.

67. *Ibid.*, p. 70.

68. *Ibid.*, Attachment C, p. 17.

69. Statement of Alfred E. Kahn, Chairman New York State Public Service Commission, on Domestic Common Carrier Regulation, to the U.S. Congress, House of Representatives, Committee on Interstate and Foreign Commerce, Subcommittee on Communications, November 18, 1975.

70. *Ibid.*

71. Additional details in: Ernest R. Kretzmer, "Communication Terminals," *Scientific American*, Vol. 227, No. 3, September 1972, pp. 133–134.

72. John K. Galbraith, *Money*, Houghton Mifflin, Boston, Ma., 1975, p. 5.

73. *The Fair Fighter*, Vol. 2, No. 2, April 1975, masthead.

74. State of New York, Communications Division, Valuation and Cost Analysis Section, *The Revenue and Cost Impact of Interconnection within the Service Area of New York Telephone Company*, State of New York, Dept. of Public Service, Albany, N.Y., January 1975, p. 99.

75. 46 FCC 2d 214 (1974), Docket No. 20003, *Notice of Inquiry*.

76. 47 U.S.C. §151, Creation of the Federal Communications Commission.

77. *National Association of Regulatory Utility Commissioners* v. *FCC*, 36 RR 2d 414 (1976).

78. *Ibid.*, footnote 98 at 414.

79. U.S. Congress. House of Representatives, 81st Congress, 1st Session. Report No. 246, "Rural Telephone Service" [to accompany H.R. 2960], March 9, 1949, pp. 2, 4.

80. U.S. Congress. House of Representatives, 92nd Congress, 2nd Session. Report No. 92–12, "Rural Telephone Bank" [to accompany H.R. 7], Feb. 22, 1971, p. 26.

81. *Ibid.*, p. 4.

82. *Ibid.*

83. *Ibid.*, p. 30.

84. *Ibid.*, p. 27.

85. *Ibid.*, p. 28.

86. *Ibid.*, p. 27.

87. These matters are discussed extensively in the filings of 29 FCC 2d 870 (1971), *First Report and Order*, Docket No. 18920, and in the hearings of the U.S. Senate committee on the Judiciary, Subcommittee on Antitrust and Monopoly, S. 1167.

88. Arthur D. Little, Inc., *The Relationship Between Market Structure and the Innovation Process*, prepared for American Telephone and Telegraph Co. by Arthur D. Little, Inc., Cambridge, Ma., January 1976, p. 1.

89. *Ibid.*, p. 2.

90. *Telecommunications Reports*, Vol. 41, No. 20, May 19, 1975, p. 7.

91. Richard Gabel, *Development of Separations Principles*, p. 148.

92. *Electronics News*, Vol. 20, No. 1029, June 2, 1975, pp. 1, 30, 31.

93. Bell Laboratories, *The Bell Labs/AT&T Relationship*, Bell Laboratories, Murray Hill, N.J., 1974, p. 11.

94. Arthur D. Little, Inc., *The Relationship Between Market Structure and the Innovation Process*, p. B–4.

95. *Ibid.*, p. B–33.

96. United System Service, Inc., response to FCC Docket No. 20003 (1974), p. 3.

97. 29 FCC 2d 915 (1971), Docket No. 18920, *First Report and Order*.

98. *Telecommunications Reports*, Vol. 41, No. 20, May 19, 1975, p. 7.

99. Order No. FCC–1043, Docket No. 19919, *Interim Decision and Memorandum Opinion and Order*, p. 14.

100. Systems Applications, Inc., *Regulatory Policy Changes and the Future*

of the Independent Telephone Industry, prepared for the U.S. Independent Telephone Association by Systems Applications, Inc., San Rafael, Calif., January 1976, p. III–30.

101. Richard Gabel, *Development of Separations Principles*, p. 31.

102. 26 FCC 2d 261 (1970), Docket No. 18866, *Recommended Report and Order of FCC–NARUC Joint Board on Jurisdictional Separations.*

103. State of New York, *The Revenue and Cost Implications of Interconnection*, p. 96.

104. 26 FCC 2d 261 (1970), Docket No. 18866.

105. *Smith* v. *Illinois Bell Telephone Company*, 282 U.S. 133 (1930).

106. William Capron (ed.), *Technological Change in Regulated Industries*, The Brookings Institution, Washington, D.C., 1971, p. 5.

107. *North Carolina Utilities Commission* v. *FCC*, ____ F. 2d ____ (4th Cir., 1976).

108. State of New York, *The Revenue and Cost Implications of Interconnection*, p. 95.

109. *Telecommunications Reports*, Vol. 39, No. 38, September 24, 1973, p. 5.

110. FCC Docket No. 20097, *Regulatory Policies Concerning Resale and Shared Use of Common Services and Facilities*, 47 FCC 2d 644 (1974), 48 FCC 2d 1077 (1974).

111. American Telephone and Telegraph Co., Response to FCC Docket No. 20097, December 11, 1974, p. 35.

112. *Ibid.*, p. 36.

113. MCI Telecommunications Corporation submission to FCC on Docket No. 20003, p. 4.

114. 47 U.S.C. §202(a), Common carriers, discriminations and preferences.

115. 47 U.S.C. §201(b), Common carriers, service and charges.

116. *Associated Press* v. *FCC*, 452 F.2d 1300, 1301 (D.C. Cir., 1971).

117. Preamble in S. 3192, H.R. 12323, H.R. 12816, H.R. 12844, H.R. 12924, and H.R. 13243 (1976).

118. 47 U.S.C. §151, Creation of the Federal Communications Commission.

119. Paul J. Berman, "Computer or Communications? Allocation of Functions and the Role of the Federal Communications Commission," *Federal Communications Bar Journal*, Vol. 27, No. 2, 1974, pp. 161–230.

120. *Baker* v. *Carr*, 369 U.S. 186 (1962).

121. John T. Dunlop, "The Limits of Legal Compulsion," pp. 1–2.

122. Richard B. Stewart, "The Reformation of American Administrative Law," *Harvard Law Review*, Vol. 88, No. 8, June 1975, p. 1813.

NOTES TO PART TWO

1. In 1972, American Telephone and Telegraph Company alone reported operating revenues of $20.9 billion, with assests of $60.6 billion. *The Fortune Double 500 Directory* 68 (1973). For the same year, General Motors operating revenues were $30.4 billion and assets were $18.2 billion. *Id.* at 4. In 1972, the

United States spent $78.3 billion for total national defense. *U.S. Budget in Brief* 78 (1973).

2. Communications Act of 1934, §1, 47 U.S.C. §151 (1970).

3. Communications Act of 1934, §§3(a)-(b), 47 U.S.C. §§153(a)-(b) (1970).

4. M.D. Flanagan, *Impact: A Compilation of Bell System Innovations In Science and Engineering Which Have Helped Create New Industries and New Products* 115–128 (1971).

5. Mann-Elkins Act, §7, ch. 309, §7, 36 Stat. 539, 544 (1910), *amending* Interstate Commerce Act of 1887, ch. 104, 24 Stat. 379 (1887).

6. Notice of Inquiry, Regulatory and Policy Problems Presented By the Interdependence of Computer and Communication Services and Facilities, 7 FCC 2d 11 (1966) [hereinafter cited as Notice of Inquiry].

7. Tentative Decision, Regulatory and Policy Problems Presented By the Interdependence of Computer and Communication Services and Facilities, 28 FCC2d 291 (1970) [hereinafter cited as Tentative Decision]; Final Decision and Order, 28 FCC2d 267 (1971) [hereinafter cited as Final Decision]; *reconsideration denied*, 34 FCC2d 557 (1972).

8. S.L. Mathison and P.M. Walker, *Computers and Telecommunications: Issues in Public Policy* 21 (1970) [hereinafter cited as Mathison & Walker].

9. Note, 116 U. Pa. L. Rev. 328, 330 (1967).

10. Note, 71 Mich. L. Rev. 172, 174 (1972).

11. M. Irwin, *The Computer Utility: Competition or Regulation?* 76 Yale L.J. 1299, 1314–1316 (1967).

12. Mathison & Walker, *supra* note 8, at 19–20.

13. D. Flanagan, *supra* note 4.

14. Notice of Inquiry, *supra* note 6, at 14.

15. Mathison & Walker, *supra* note 8, at 21.

16. Notice of Inquiry, *supra* note 6.

17. *Id.*

18. Tentative Decision, *supra* note 7, at 295.

19. Mathison & Walker, *supra* note 8, at 34.

20. Note, 71 Mich. L. Rev., *supra* note 10, at 188.

21. *Id.* at 189–194.

22. Final Decision, *supra* note 7, at 270.

23. *Id.* at 268–275.

24. 47 C.F.R. §64.702(c) (1972), *as amended* 38 Fed. Reg. 8744 (1973).

25. United States v. Western Electric Co., 1956 Trade Cas. ¶68,246 at 71, 134, 13 P & F Radio Reg. 2143 (D.C.N.J. 1956).

26. 474 F.2d 724 (2d. Cir. 1973).

27. *Id.* at 736.

28. Tentative Decision, *supra* note 7, at 298.

29. *Id.* at 297.

30. *Id.* at 298; Final Decision, *supra* note 7, at 270.

31. 47 C.F.R. §64.702(a)(2) (1973).

32. 47 C.F.R. §64.702(a)(1) (1973).

33. A comparison of the definition of message switching and data processing illustrates the difficulty of basing functional distinctions on descriptions of the operations employed to accomplish those functions. The definitions of data processing and message switching, if taken together, seem to imply that a data processor alters the content of the information that is processed. Yet it is entirely plausible to expect that certain information retrieval services will base their operations on the ability of one person to store data and another to find and retrieve it, unaltered in content.

Furthermore, it is clear that computer controlled transmission of messages involves some data processing. The FCC noted this in the Tentative Decision, 28 FCC2d 291, 296, n. 1: "in message-switching, any information transmitted from terminal points must necessarily pass through and undergo some processing by the computer mainframe." Thus, it does not seem clear whether the FCC is basing its distinction on the operations being performed or on the reasons for which one offers or purchases the ability to perform the operations.

34. Final Decision, *supra* note 7, at 278.

35. Tentative Decision, *supra* note 7, at 304.

36. 47 C.F.R. §§64.702(e)-(f) (1973).

37. Tentative Decision, *supra* note 7, at 305.

38. Mathison & Walker, *supra* note 8, at 63.

39. Mann-Elkins Act, §7, ch. 309, §7, 36 Stat. 539, 544 (1910), *amending* Interstate Commerce Act of 1887, ch. 104, 24 Stat. 379 (1887).

40. *See* discussion of communications common carriage, pp. 174–187 *infra*.

41. Final Decision, *supra* note 7, at 279–80.

42. The specific issue here is the jurisdiction of an administrative agency to determine its own jurisdiction. "Much the most difficult problems of the scope of review of administrative determinations are . . . presented in [the questions] . . . [w]hat is, or what should be the scope of judicial review of (a) an agency's application of a statutory term to undisputed facts . . . or (b) an agency's interpretation of a statutory term?" W. Gellhorn & C. Byse, *Administrative Law, Cases and Comments* 379 (5th ed., 1970). *See also* 4 Davis, *Admin. Law Treatise* ch. 30 (1958); L.L. Jaffe, *Judicial Control of Administrative Action* chs. 14–16 (1965).

43. Communications Act of 1934, §205, 47 U.S.C. §205 (1970).

44. Mathison & Walker, *supra* note 8, at 64.

45. M. Irwin, *Time-Shared Information Systems: Market Entry in Search of A Policy*, 31 Proceedings of the 1967 Fall Joint Computer Conference 513, 518 (1967).

46. Interview with Professor Hendrik Bode, Division of Engineering and Applied Physics, Harvard University, Cambridge, Massachusetts, August 1973.

47. Telephony, September 17, 1973, at 36.

48. *New York Telephone Company—Directory Assistance Charges*, Order and Opinion Approving New York Telephone Company's Institution of Directory Assistance Charges, New York State Public Service Commission, Opinion No. 74–19 (Case 26272, May 31, 1974) [hereinafter cited as New York Public Service Comm'n Opinion and Order].

49. Brief of New York Telephone Company, *New York Telephone Company—Directory Assistance Charges* (N.Y. State Pub. Service Comm'n Case

26272, 1973) at 8–9; Brief of the Public Service Comm'n Staff, *New York Telephone Company—Directory Assistance Charges* (N.Y. State Pub. Service Comm'n Case 26272, 1973) at 5–6 [hereinafter cited as Staff Brief].

50. Staff Brief, *supra* note 49, at 17–18.

51. New York Public Service Comm'n Opinion and Order, *supra* note 48 at 9.

52. *Id.* at 16. The 16.5¢ cost of each call to directory assistance is comprised of 11.3¢ operating costs and 5.2¢ fixed costs per call. Thus, in not allocating costs fully, the plan does not fully succeed in distributing the costs of the service to the subscribers responsible for them, but does allow NYT to offer the identifiable service of directory assistance at a rate far below its own costs.

53. *Id.* at 15. The rationale given for the decision to charge for calls to directory assistance when the number is listed in any NYT directory is that when a subscriber frequently seeks numbers contained in nonlocal directories, he will be able to obtain the pertinent directory from NYT. *Id.* at 14.

54. *Id.* at 25.

55. *Id.* at 17.

56. *Id.* at 20; at 23, the PSC suggested that the increased costs of directory assistance due to use of the service because unpublished numbers are not listed in directories would be borne by those requesting unpublished numbers at some later time.

57. *Id.* at 5.

58. *Id.* at 4.

59. *Id.* at 27.

60. *See* discussion of Automatic Location Identification, pp. 165–169, *infra.*

61. F.R. Stroud, *A Systems Analysis of the Single Emergency Telephone Number and Automatic Location Identification* 7 June 1972 (General Telephone and Electronics).

62. *Id.*

63. L.A. Ceddia, "*911,*" *A Staff Regulatory Viewpoint of the Universal Emergency Telephone Number* 9–10, May 15, 1973 (Communications Division, Dep't of Public Service, State of New York).

64. F.R. Stroud, *supra* note 61, at 27.

65. Telecommunications Reports, December 3, 1973, at 24.

66. F.R. Stroud, *supra* note 61, at 8.

67. *Id.* at 19–20.

68. *Id.* at 67–70.

69. *Id.* at 70–74.

70. In fact, there does not seem to be a reason why each local jurisdiction could not install its own on-line ALI system, without using "911" and without using the telephone company for data processing services at all.

71. F.R. Stroud, *supra* note 61, at 93.

72. Letter from Mr. Bernard Strassburg, Chief, Common Carrier Bureau, Federal Communications Commission, to Mr. William Malone, Counsel, General Telephone and Electronics, August 18, 1972.

73. Use of the Carterfone in Message Toll Telephone Service, 13 FCC2d 420 (1968), *reconsideration denied*, 14 FCC2d 571 (1968).

74. F.R. Stroud, *supra* note 61, at 70.

75. *Id.* at 73.

76. *Id.* at 70.

77. K.R. Bordner & J.S. Huston, *911: A Study of the Single Emergency Telephone Number* 26, March, 1970 (Benj. Franklin Research Laboratories, Philadelphia, Pennsylvania).

78. Southwestern Bell Telephone Co. v. Dialite Dial Co., 102 F. Supp. 872 (W.D. Okla., 1951).

79. Application of Packet Communications, Inc., FCC File No. P–C–8533, (1973).

80. Memorandum Opinion, Order and Certificate, 43 FCC2d 922 (1973).

81. *Id.* at 926.

82. Further technical explanation can be found in R.M. Metcalfe, Packet Communication, April, 1973 (unpublished Ph.D. Thesis, Harvard University, Cambridge, Massachusetts). Many packet switching techniques were developed in the experimental computer network of the Advanced Research Projects Agency, Department of Defense.

83. L.R. Talbert, *PCI: A Complete Data Communication Service*, March 12, 1973 (19th Annual Management Conference, American Management Association, New York).

84. Application of Packet Communications, Inc., *supra* note 79, at 4; Comments of the MCI Carriers at 2, FCC File No. P–C–8533, (1973).

85. Comments of Western Union, FCC File No. P–C–8533, (1973).

86. Comments of the MCI Carriers, *supra* note 84.

87. Comments of Datran, FCC File No. P–C–8533, (1973).

88. 6 FCC 562 (1938).

89. *Mackay* articulates a policy that does not seem to be fully applicable to the application of PCI. First, Mackay Radio and Telegraph Company was leasing lines of another company between Baltimore and Washington with the intent of duplicating a public telegraph service. Mackay's enterprise was not based on a new service or on a new and rapidly developing technology, at the time. Mackay itself was an old organization which was, at the time of the proceeding, already a common carrier subject to the powers of the FCC.

Second, the decision in *Mackay* was based on a strong analogy of communications lines to railroad tracks. This analogy may not be entirely sound, as will be discussed pages 174–176.

Finally, the Commission also based its decision in *Mackay* on the argument that allowing unlimited resale of lines would be ruinous for existing carriers in terms of the competition that would result in geographic areas that the FCC has assigned to a limited number of carriers. It is difficult to see the force of this argument, since offerings such as PCI would be leasing lines from already established carriers, and any increased demand for communications services would also accrue to those carriers, hardly setting up ruinous competition. It is, of course, possible that PCI's service, because of its efficiency, will result in a decrease in demand for the established common carriers (see pages 188–189, *infra.*). But whether or not this possibility and its results should be studied *before* a new enterprise is allowed to enter the market demands yet another policy decision. Policy makers must decide whether to favor innovation and implementation of new services without any barriers to market entry or whether to slow such implementation and innovation through careful study of its effects on a plausibly adequate communications market status quo.

90. Comments of the MCI Carriers, *supra* note 84, at 6.

91. Reply Comments of Packet Communications, Inc. at 5, FCC File No. P–C–8533 (1973).

92. Application of Packet Communications, Inc., *supra* note 79, at 4.

93. American Telephone & Telegraph, FCC Application No. 931, February 26, 1973.

94. Memorandum Opinion, Order and Certificate, 43 FCC2d 922, 925 (1973).

95. 47 U.S.C. §153(h) (1970).

96. Interview with Professor Hendrik Bode, Division of Engineering and Applied Physics, Harvard University, Cambridge, Massachusetts, August 1973.

97. See text accompanying note 32, *supra*.

98. The Propeller Niagara v. Cordes, 21 How. (62 U.S.) 7, 22 (1858); Washington *ex rel.* Stimson v. Kuykendall, 275 U.S. 207, 211 (1927).

99. Gardner v. Boston Elevated Railway, 204 Mass. 213, 216, 90 N.E. 534, 535 (1910).

100. Kettendorfer v. Globe Transfer and Storage Co., 70 Wash. 645, 127 P. 295 (1912).

101. 154 U.S. 1 (1893).

102. *Id.* at 14.

103. *Id.*

104. Use of the Carterfone in Message Toll Telephone Service, 13 FCC2d 420 (1968).

105. Application of Packet Communications, Inc., *supra* note 79, Illustrative Tariff Rates and Regulations at 18.

106. Mann-Elkins Act, §7, ch. 309, §7, 36 Stat. 539, 544 (1910).

107. Interstate Commerce Act of 1887, ch. 104, 24 Stat. 379 (1887).

108. 45 Cong. Rec. 6975 (1910) (remarks of Senator Elkins).

109. 45 Cong. Rec. 5537 (1910) (remarks of Representative Mann).

110. 45 Cong. Rec. 5533 (1910) (remarks of Representative Mann).

111. Act to Regulate Radio Communications, ch. 287, 37 Stat. 302 (1910).

112. Radio Act of 1927, ch. 169, 44 Stat. 1162 (1927).

113. Communications Act of 1934, §3(h), 47 U.S.C. §153(h) (1970).

114. H.R. Conf. Rep. No. 1918, 73d Cong., 2d Sess. 46 (1934).

115. Note, 116 U. Pa. L. Rev., *supra* note 9, at 338.

116. 24 FCC 251, 16 P & F Radio Reg. 1005 (1958).

117. *Id.* at 254, 16 P & F Radio Reg. at 1008–9.

118. *See* text accompanying note 98, *supra*.

119. Note, 116 U. Pa. L. Rev., *supra* note 9, at 338.

120. Communications Act of 1934, §1,47 U.S.C. §151 (1970).

121. *E.g.*, National Broadcasting Co. v. United States, 319 U.S. 190 (1943); United States v. Southwestern Cable Co., 392 U.S. 157 (1968).

122. Communications Act of 1934, §2(a), 47 U.S.C. §152(a) (1970).

123. The exception, §223, 47 U.S.C. §223 (1970), deals with making obscene or harassing phone calls.

124. 47 U.S.C. §§153(a)-(b) (1970).

125. Communications Act of 1934, §214(a), *as amended*, 47 U.S.C. §214 (a) (1970).

126. See pages 181–183, *infra.*

127. Interview with Professor Hendrik Bode, Division of Engineering and Applied Physics, Harvard University, Cambridge, Massachusetts, August 1973.

128. Telephone interview with Mr. Kelley E. Griffith, Chief, Domestic Rates Division, Common Carrier Bureau, Federal Communications Commission, July 1973.

129. Communications Act of 1934, §§2(b)(2)-(4), 47 U.S.C. §§152(b)(2)-(4) (1970).

130. " 'Connecting carrier' means *a carrier* described in clauses (2), (3), or (4) of section 152(b) of this title." 47 U.S.C. §153(u) (1970) (emphasis added).

131. Kitchen v. Federal Communications Commission, 464 F.2d 801 (D.C. Cir. 1972).

132. *Id.* at 803, n. 5.

133. Final Decision, *supra* note 7, at 286.

134. I.T.T. World Communications, Inc., Rates, Rules and Regulations Applicable to Customer Data Retransmission Service, FCC No. 65, Original Page 5, May 7, 1973, Effective June 7, 1973.

135. 45 Cong. Rec. 6973 (1910) (remarks of Senator Dixon).

136. 45 Cong. Rec. 5533 (1910) (remarks of Representative Bartlett).

137. 78 Cong. Rec. 4139 (1934) (remarks of Senator Dill).

138. *See* text accompanying notes 136 and 29, *supra.*

139. Mathison & Walker, *supra* note 8, at 68; Letter from Dr. Clay T. Whitehead, Director, Office of Telecommunications Policy, Executive Office of the President, to Mr. Dean Burch, Chairman, Federal Communications Commission, January 24, 1974.

140. *See* text accompanying note 135, *supra.*

141. H.R. Conf. Rep. 1918, 73d Cong., 2d Sess. 45–46 (1933). Associated Press's membership restrictions were enjoined in an antitrust action, *Associated Press v. United States*, 326 U.S. 1 (1945). Yet even though this decision might be taken to require press services to serve all comers, these services still are not subject to FCC jurisdiction. Press services provide an important information distribution function. The fact that they are prohibited from having restrictive membership or customer rules as a matter of antitrust, not communications law indicates that the law and regulations of the FCC need not be the exclusive means for deciding the availability of information distribution services and functions.

142. Interview with Professor Richard Stewart, Harvard Law School, Cambridge, Massachusetts, August 1973.

143. *See* note 42, *supra.*

144. *E.g.*, Frontier Broadcasting Co. v. Collier, 24 FCC 251 (1958); T.V. Repeater Services, 26 FCC 403, 431 (1959).

145. Notice of Inquiry and Proposed Rule-Making, Distribution of Television Broadcasting By Community Antenna Television Systems, 30 Fed. Reg. 6072 6078 (1965); *See also* S. 2653, 86th Cong., 1st Sess; S. Rep. No. 923, 86th Cong. 1st Sess. (1960).

146. Second Report and Order, Community Antenna Television Services, 2 FCC2d 725 (1966).

147. Southwestern Cable Co. v. United States, 378 F.2d 118 (9th Cir. 1967), *rev'd* 392 U.S. 157 (1968).

148. Professor Stewart notes that with respect to CATV, the Courts gave considerable weight to the FCC's finding that CATV was regulatable as "ancillary to broadcasting." This deference need not necessarily be afforded to FCC findings as to services which rely on converging computer and communications technologies. The *GTE* case provides an example of a court's willingness to "fence in" the FCC.

149. Philadelphia Television Broadcasting Co. v. Federal Communications Commission, 123 U.S. App. D.C. 298, 359 F.2d 282 (1966).

150. Note, 116 U. Pa. L. Rev., *supra* note 9, at 337.

151. Memorandum Opinion, Order and Certificate, 43 FCC2d 922, 925 (1973).

152. Packet proposes to sell its service solely on the basis of the amount of data a subscriber sends. Packet assumedly will pay for channel capacity even though there might not be any paying subscribers; carrier channels are available only on a relatively long term basis. Thus, in order to be profitable, there must be a threshold amount of demand for Packet's service before it can afford to obtain a channel from the carriers.

153. *E.g.*, 18 U.S.C. §1304 (1970) (broadcasting lottery information); 18 U.S.C. §1343 (1973) (perpetrating fraud by wire, radio or television); 47 U.S.C. §223 (1970) (making obscene or harassing phone calls).

154. Mathison & Walker, *supra* note 8, at 138.

155. Telecommunications Reports, March 25, 1974, at 1. The effective date of the tariff was extended from May 18 until July 18, 1974, after the Datran Corp. complained to the FCC that it would take at least four months to evaluate the 2300 pages of material that AT&T filed in support of the tariff. Telecommunications Reports, April 15, 1974, at 10. Effective date of the tariff was subsequently postponed until September 5, 1974, at the request of the FCC. Telecommunications Reports, July 22, 1974, at 14.

156. Mathison & Walker, *supra* note 8, at 139–140.

157. *Id.* at 135, citing Western Union Petition *re* Revisions of AT&T Tariff No. 260 (private line services), January 20, 1969, at 24.

158. Note, 71 Mich. L. Rev., *supra* note 10, at 184, n. 60.

159. For example, the issue of allowing private or non–telephone company equipment to be interconnected with the public telephone network has been raised anew on the state level. *See* Telecommunications Reports, September 3 and 10, 1973. The issue ostensibly had been settled by the FCC's *Carterfone* ruling, 13 FCC2d 420 (1968). The FCC ruled again, Telerent Leasing Corp., 45 FCC2d 204 (1974), holding that the FCC, not a state agency, has primacy of authority over interconnection with the nationwide telephone network. The

FCC decision barring state regulation on the interconnection issue is being appealed by the North Carolina Utilities Commission, North Carolina Utilities Commission v. United States, Case No. 74–1220 (4th Cir., filed February 27, 1974).

160. Final Decision, *supra* note 7, at 278.

161. G.T.E. Service Corp. v. Federal Communications Commission, 474 F.2d 724 (2d Cir. 1973).

162. Communications Act of 1934, §§2(b)(2)-(4), 47 U.S.C. §§152(b)(2)-(4) (1970).

163. 47 U.S.C. §§201–205 (1970).

164. 47 U.S.C. §214 (1970).

165. United States v. Western Electric Co., 1956 Trade Cas. ¶68,246 at 71, 134, 13 P & F Radio Reg. 2143 (D.C.N.J. 1956).

NOTES TO PART THREE

1. Sources:
   a) Statistics on U.S. Foreign Trade provided by The U.S. Bureau of Census, Foreign Trade Division, Washington, D.C., by telephone, November 19, 1975.
   b) Statistics on Public Telephone and Telegraph and Private Circuit Telephone and Telegraph from *Statistics of Communications Common Carriers, Year Ended December 31, 1973*, U.S. Federal Communications Commission, pages 23, 167, U.S. Government Printing Office, Washington, D.C.
   c) Statistics on Non-Governmental International Organizations ("International Affinity Groups") from Union of International Associations Yearbook (12th ed.), p. 13, Brussels, 1969.
   d) Overseas messages from American Telephone and Telegraph Co., *Bell System Statistical Manual* (Murray Hill, N.J.: Bell Laboratories) section 800 "Volume of Communications," p. 807.
   e) Statistics on airline messages provided to the author by Hal Burt, Deputy Manager, North America, Société Internationale De Telecommunications Aeronautiques.

2. See Alex Inkeles, "The Emerging Social Structure of the World," presented at the International Political Science Association meetings at Montreal, Canada, August 20–23, 1973.

3. Samuel P. Huntington, "Transnational Organizations in World Politics," *World Politics*, Vol. XXV, No. 3, April 1973, p. 342.

4. The International Telecommunications Union, the oldest of the United Nations specialized agencies (founded in 1865) produces the mass of its highly detailed reports and recommendations through its International Radio Consultative Committee (CCIR) and International Telegraph and Telephone Consultative Committee (CCITT).

5. See the Committee on the Organization of the Government for the Conduct of Foreign Policy Report, esp. Ch. 10, "Planning for Foreign Policy," pp. 143–150, U.S. Government Printing Office, Washington, D.C., June 1975.

6. The White House, President's Science Advisory Committee, "The World Food Problem," Vols. I and II, May 1967, Vol. III, September 1967, U.S. Government Printing Office, Washington, D.C.

7. Joseph S. Nye, Jr. and Robert O. Keohane, "An Introduction," *Transnational Relations and World Politics*, Nye and Keohane (eds.), Harvard University Press, Cambridge, Mass., 1972, p. 1.

8. Huntington, *op. cit.*, p. 333.

9. J.J. Servan-Schreiber, *The American Challenge*, Atheneum, New York, 1968, p. 101.

10. *United States Treaties and Other International Agreements Pertaining to Telecommunications*, OT Report 74–26, U.S. Department of Commerce, Office of Telecommunications, U.S. Government Printing Office, Washington, D.C., January 1974, has 73 entries under the heading "Defense."

11. *Communications News*, September 1975, p. 3.

12. The comment was made to the author during an interview with an executive of a U.S. international record carrier in August, 1975.

13. More than 30 individuals were interviewed for this study, among them were twelve communications managers of transnational organizations.

14. *Ibid.*

15. *Ibid.*

16. GE Information Service, a General Electric publication, October 1974, no. 0910.00C.

17. *8th Annual Industry Report to ADAPSO*, 1974, Quantum Science Corporation, New York, made available to the author by the Association of Data Processing Organization, Inc., Montvale, N.J.

18. *Ibid.*

19. The examples were provided to the author with an understanding that the companies would not be explicitedly identified.

20. *Proposals for a Communications Policy for Canada*, Information Canada, Ottawa, 1973, p. 11.

21. In conjunction with the reopening of FCC Docket 18875 on facility planning a "Traffic Forecasting Working Group" was formed. The quotation is from its report of May 8, 1975, a copy of which was made available to the author.

Index

About the Authors

Anthony G. Oettinger, director of the Harvard Program on Information Resources Policy, is Professor of Information Resources Policy, a member of the Faculty of Public Administration and Gordon McKay Professor of Applied Mathematics at Harvard University.

He is chairman of the CATV Commission of the Commonwealth of Massachusetts and a consultant to the National Security Council, Executive Office of the President of the United States. He is a member of the Research Advisory Board of the Committee for Economic Development. Since 1956, he has served as a consultant to Arthur D. Little, Inc., on the uses of information technologies in many industries.

He has been president of the Association for Computing Machinery (1966–68) and a consultant to the Office of Science and Technology, Executive Office of the President of the United States (1961–73). He is a Fellow of the American Academy of Arts and Sciences, the American Association for the Advancement of Science and the Institute of Electrical and Electronics Engineers.

As chairman of the Computer Science and Engineering Board of the National Academy of Sciences (1967–73), he led the preparation of numerous NAS reports, including *A Technical Analysis of the Common Carrier/Use Interconnections Area* (Lewis S. Billig, Project Director), and *Databanks in a Free Society: Computers, Record Keeping and Privacy* (Alan F. Westin, Project Director).

Paul J. Berman is associated with the Washington, D.C. law firm of Covington & Burling, and practices predominantly in the area of communications law. A graduate in computer science from Harvard College, he continued his studies of the impact of developing information technologies at the Harvard Law School. While associated with the Harvard Program on Information Resources Policy, he authored "Computer or Communications" and co-authored "The Medium and the Telephone" with the Program's Director, Professor Anthony G. Oettinger. Mr. Berman's article, "CATV Leased Access Channels and the FCC—The Intractable Jurisdiction Question," appeared in the November, 1975 issue of the Notre Dame Lawyer.

William H. Read began working on information policy at Harvard in 1974 while on leave from the U.S. Information Agency. He initially did research on the international flow of mass media, with the findings published in *America's Mass Media Merchants.* His interest in information research, which now includes telecommunications, is rooted in a career of practical experience as a journalist and a foreign service information officer.

List of Contributors

Harvard Program on Information Resources Policy

AFFILIATES

American Can Company
American Telephone and Telegraph
Arthur D. Little Foundation
Bell Canada
Booth Newspapers
Bunker Ramo Corporation
Codex Corporation
Communications Workers of America
Computer and Communications
  Industry Association
Data Transmission Company
Donaldson, Lufkin & Jenrette
L.M. Ericsson (Sweden)
Federal Communications Commission
Federal Reserve Bank of Boston
First National Bank of Boston
First National Bank of Chicago
General Electric Company
Harte-Hanks Newspapers
IBM Corporation
International Data Corporation

*As of December, 1976.

International Paper Company
International Resources Development, Inc.
Lee Enterprises
Lockheed Missiles and Space Company
John & Mary R. Markle Foundation
Mead Corporation
Minneapolis Star and Tribune Company
New York Times Company
Nippon Electric Company
Norfolk & Western Railway Company
Payment Systems, Inc.
Pitney Bowes, Inc.
PRC Information Sciences Company
Rockefeller Brothers Fund
Rockwell International
Salomon Brothers
Seiden & De Cuevas, Inc.
Systems Applications, Inc.
The Boston Globe
Time Incorporated
Transamerica Corporation
United Telecommunications, Inc.
U.S. Department of Commerce:
  National Technical Information Service
  Office of Telecommunications
Western Union Corporation
Western Union International, Inc.
Zenith Radio Corporation

STUDENT ASSISTANTSHIPS

William Bingham Foundation
Bank of America
Chase Manhattan Bank